D. B. Cheever
44 Washington St
Ann Arbor

U of M
Lit 91.

Feb 21st 89

ELEMENTS

OF

DESCRIPTIVE GEOMETRY,

WITH ITS APPLICATIONS TO

SPHERICAL PROJECTIONS, SHADES AND SHADOWS, PERSPECTIVE AND ISOMETRIC PROJECTIONS.

BY

ALBERT E. CHURCH, LL. D.,

PROFESSOR OF MATHEMATICS IN THE U. S. MILITARY ACADEMY, AUTHOR OF ELEMENTS OF THE DIFFERENTIAL AND INTEGRAL CALCULUS, AND OF ELEMENTS OF ANALYTICAL GEOMETRY.

A. S. BARNES & COMPANY,

NEW YORK AND CHICAGO.

Valuable Works by Leading Authors

IN THE

HIGHER MATHEMATICS.

A. E. CHURCH, LL.D.,

Prof. Mathematics in the U. S. Military Academy, West Point.

CHURCH'S ANALYTICAL GEOMETRY.
CHURCH'S CALCULUS.
CHURCH'S DESCRIPTIVE GEOMETRY

EDWARD M. COURTENAY, LL.D.,

Late Prof. Mathematics in the University of Virginia.

COURTENAY'S CALCULUS.

CHAS. W. HACKLEY, S. T. D.,

Late Prof. of Mathematics and Astronomy in Columbia College.

HACKLEY'S TRIGONOMETRY.

W. H. C. BARTLETT, LL.D.,

Prof. of Nat. & Exp. Philos. in the U. S. Military Acad., West Point

BARTLETT'S SYNTHETIC MECHANICS.
BARTLETT'S ANALYTICAL MECHANICS.
BARTLETT'S ACOUSTICS AND OPTICS.
BARTLETT'S ASTRONOMY.

DAVIES & PECK,

Department of Mathematics, Columbia College.

MATHEMATICAL DICTIONARY.

CHARLES DAVIES, LL.D.,

Late of the United States Military Academy and of Columbia College

A COMPLETE COURSE IN MATHEMATICS

See A. S. Barnes & Co.'s Descriptive Catalogue.

C. D. G.

PREFACE.

THESE pages have been written and are published with the single object of presenting, in proper form to be used as a text-book, the course of Descriptive Geometry, as taught at the U. S. Military Academy.

Without any effort to enlarge or originate, the author has striven to give, with a natural arrangement and in clear and concise language, the elementary principles and propositions of this branch of science, of so much interest to the mathematical student, and so necessary to both the civil and military engineer.

Though indebted for many of the ideas to the early instructions of his predecessor and friend, Professor Davies, whose text-books on this subject were among the first in the English language, the author has been much aided by a frequent reference to the French works of Leroy and Olivier, and to the elaborate American work of Professor Warren.

It is intended to include, in an edition to be issued at an early day, the application of the subject to shades and shadows, and perspective.

U. S. MILITARY ACADEMY,
October, 1864

CONTENTS.

PART I.

ORTHOGRAPHIC PROJECTION.

PART II.

SPHERICAL PROJECTIONS.

PART III.

SHADES AND SHADOWS.

PART IV.

LINEAR PERSPECTIVE.

PART V.

ISOMETRIC PROJECTIONS.

PART I.

ORTHOGRAPHIC PROJECTIONS.

PRELIMINARY DEFINITIONS.

1. DESCRIPTIVE GEOMETRY is that branch of Mathematics which has for its object the explanation of the methods of representing by drawings:

First. All geometrical magnitudes.

Second. The solution of problems relating to these magnitudes in space.

These drawings are so made as to present to the eye, situated at a particular point, the same appearance as the magnitude or object itself, were it placed in the proper position.

The representations thus made are *the projections of the magnitude or object.*

The planes upon which these projections are usually made are *the planes of projection.*

The point, at which the eye is situated, is *the point of sight.*

2. When the point of sight is in a perpendicular, drawn to the plane of projection, through any point of the drawing, and at an infinite distance from this plane, *the projections are Orthographic.*

When the point of sight is within a finite distance of the drawing, *the projections are Scenographic*, commonly called *the Perspective* of the magnitude or object

3. It is manifest that, *if a straight line be drawn through a given point and the point of sight, the point, in which this line pierces the plane of projection, will present to the eye the same appearance as the point itself, and therefore be the projection of the point on this plane.*

The line thus drawn is *the projecting line of the point.*

4. In the *Orthographic Projection*, since the point of sight is at an infinite distance, the projecting lines drawn from any points of an object, of finite magnitude, to this point, *will be parallel to each other and perpendicular to the plane of projection.*

In this projection two planes are used, at right angles to each other, the one *horizontal* and the other *vertical*, called respectively *the horizontal* and *vertical plane of projection.*

5. In Fig. 1, let the planes represented by ABF′ and BAD be the two planes of projection, the first *the horizontal* and the second *the vertical.*

Their line of intersection AB is *the ground line.*

These planes form by their intersection four diedral angles. The *first angle*, in which the point of sight is always situated, is above the horizontal and in front of the vertical plane. *The second* is above the horizontal and behind the vertical. *The third* is below the horizontal and behind the vertical. *The fourth*, below the horizontal and in front of the vertical, as marked in the figure.

REPRESENTATION OF POINTS.

6. Let M, Fig. 1, be any point in space. Through it draw

Mm perpendicular to the horizontal, and Mm' perpendicular to the vertical plane; m will be the projection of M on the horizontal, and m' that on the vertical plane, Art. (4). Hence, *the horizontal projection of a point is the foot of a perpendicular through the point, to the horizontal plane; and the vertical projection of a point is the foot of a perpendicular through it to the vertical plane.*

The lines Mm and Mm' are the horizontal and vertical projecting lines of the point.

7. Through the lines Mm and Mm' pass a plane. It will be perpendicular to both planes of projection, since it contains a right line perpendicular to each, and therefore perpendicular to the ground line AB. It intersects these planes in the lines mo and $m'o$, both perpendicular to AB at the same point, forming the rectangle Mo. By an inspection of the figure it is seen that

$$Mm = m'o, \quad \text{and} \quad Mm' = mo;$$

that is, the distance of the point M, from the horizontal plane, *is equal to the distance of its vertical projection from the ground line;* and the distance of the point from the vertical plane is equal to that of its horizontal projection from the ground line.

8. *If the two projections of a point are given, the point is completely determined;* for if at the horizontal projection m, a perpendicular be erected to the horizontal plane, it will contain the point M. A perpendicular to the vertical plane at m', will also contain M; hence the point M is determined by the intersection of these two perpendiculars.

If M be in the horizontal plane, Mm = 0, and the point is its own horizontal projection. The vertical projection will be in the ground line at o.

If M be in the vertical plane, it will be its own vertical projection, and its horizontal projection will be in the ground line at o.

If M be in the ground line, it will be its own horizontal, and also its own vertical projection.

REPRESENTATION OF PLANES.

9. Let tTt', Fig. 2, be a plane, oblique to the ground line, intersecting the planes of projection in the lines tT and $t'T$ respectively. It will be completely determined in position by its two lines tT and $t'T$.

Its intersection with the horizontal plane is *the horizontal trace* of the plane, and its intersection with the vertical plane is *the vertical trace.* Hence, a plane is given by its traces.

Neither trace of this plane can be parallel to the ground line; for if it should be, the plane would be parallel to the ground line, which is contrary to the hypothesis.

The two traces must intersect the ground line at the same point, for if they intersect it at different points, the plane would intersect it in two points, which is impossible.

If the plane be parallel to the ground line, as in the same figure, *its traces must be parallel to the ground line;* for if they are not parallel they must intersect it; in which case the plane would have at least one point in common with the ground line, which is contrary to the hypothesis.

If the plane be parallel to either plane of projection, it will have but one trace, which will be on the other plane and parallel to the ground line.

10. If the given plane be perpendicular to the horizontal plane, *its vertical trace will be perpendicular to the ground line,* as $t''T$, in Fig. 2; for the vertical plane is also perpendicular to the horizontal plane; hence the intersection of the two planes, which *is the vertical trace,* must be perpendicular to the hor-

zontal plane, and therefore to the ground line which passes through its foot.

Likewise, if a plane be perpendicular to the vertical plane, its horizontal trace will be perpendicular to the ground line.

If the plane simply pass through the ground line, its position is not determined.

If two planes are parallel, their traces on the same plane of projection are parallel, for these traces are the intersections of the parallel planes by a third plane.

REPRESENTATION OF RIGHT LINES.

11. Let MN, Fig. 3, be any right line in space. Through it pass a plane Mmn perpendicular to the horizontal plane; mn will be its horizontal, and pp', perpendicular to AB, Art. (10), its vertical trace. Also through MN pass a plane M$m'n'$, perpendicular to the vertical plane: $m'n'$ will be its vertical, and $o'o$ its horizontal trace. The traces mn and $m'n'$ are the projections of the line. Hence, *the horizontal projection* of a right line *is the horizontal trace of a plane passed through the line perpendicular to the horizontal plane;* and *the vertical projection* of a right line *is the vertical trace of a plane, through the line perpendicular to the vertical plane.*

The planes Mmn and M$m'n'$ are respectively *the horizontal* and *vertical* projecting planes of the line.

12. The two projections of the line being given, the line will, in general, be completely determined; for if through the horizontal projection we pass a plane perpendicular to the horizontal plane, it will contain the line; and if through the vertical projection we pass a plane perpendicular to the vertical plane, it will also contain the line. The intersection of these planes must, therefore, be the line. Hence we say, *a right line is given by its projections.*

13. The projections mn and $m'n'$ are also manifestly made up of the projections of all the points of the line MN. Hence, if a right line pass through a point in space, *its projections will pass through the projections of the point.* Likewise, if any two points in space be joined by a right line, the projections of this line will be the right lines joining the projections of the points on the same plane.

14. If the right line be perpendicular to either plane of projection, its projection on that plane will be a point, and its projection on the other plane will be perpendicular to the ground line. Thus in Fig. 4, Mm is perpendicular to the horizontal plane: m is its horizontal, and $m'o$ its vertical projection.

If the line be parallel to either plane of projection, its projection on that plane will evidently be *parallel to the line itself*, and its projection on the other plane *will be parallel to the ground line.* For the plane which projects it on the second plane must be parallel to the first; its trace must therefore be parallel to the ground line, Art. (9); but this trace is the projection of the line. Thus MN is parallel to the horizontal plane, and $m'n'$ is parallel to AB. Also, the definite portion MN, of the line, *is equal to its projection mn*, since they are opposite sides of the rectangle Mn.

If the line is parallel to both planes of projection, or to the ground line, both projections will be parallel to the ground line.

If the line lie in either plane of projection, its projection on that plane will be the line itself, and its projection on the other plane will be the ground line. Thus in Fig. 5, MN, in the vertical plane, is its own vertical projection, and mn or AB is its horizontal projection.

15. If the two projections of a right line are perpendicular to the ground line, the line is undetermined, as the two pro-

jecting planes coincide, forming only one plane, and do not by their intersection determine the line as in Art. (12).

All lines in this plane will have the same projections. Thus in Fig. 5, *mn* and *m'n'* are both perpendicular to AB; and any line in the plane MN*o* will have these for its projections.

If, however, the projections of two points of the line are given, the line will then be determined; that is, if *mm'* and *nn'* are given, the two points M and N will be determined, and, of course, the right line which joins them.

All lines and points, situated in a plane perpendicular to either plane of projection, will be projected on this plane in the corresponding trace of the plane.

16. If two right lines are parallel, *their projections on the same plane will be parallel.* For their projecting planes are parallel, since they contain parallel lines and are perpendicular to the same plane; hence their traces will be parallel, Art. (10); but these traces are the projections

REVOLUTION OF OBJECTS.

17. Any geometrical magnitude or object is said to be revolved about a right line as an axis, *when it is so moved that each of its points describes the circumference of a circle whose plane is perpendicular to the axis, and whose centre is in the axis.*

By this revolution, it is evident that the relative position of the points of the object is not changed, each point remaining at the same distance from any of the other points. Thus, if the point M, Fig. 6, be revolved about an axis DD', in the horizontal plane, it will describe the circumference of a circle whose centre is at *o* and whose radius is M*o*; and since the point must remain in the plane perpendicular to DD', when it reaches the horizontal plane it will be at *p* or *p'*, in the perpendicular *mop*, at a distance from *o* equal to M*o*; that is, it will be found *in a straight line passing through its horizontal projection perpendicular to the axis*, and at a distance from the axis equal *to the hypothenuse of a right-angled triangle of which the base* (*mo*) *is the distance from the horizontal projection to the axis, and the altitude* (M*m*) *equal to the distance of the point from the horizontal plane, or equal to the distance* (*m'r*) *of its vertical projection from the ground line.*

Likewise, if a point be revolved about an axis in the vertical plane until it reaches the vertical plane, its revolved position will be found by the same rule, changing the word *horizontal* into *vertical*, and the reverse.

If the axis pass through the horizontal projection of the point, in the first case, the base of the triangle will be 0; the hypothenuse becomes equal to the altitude, and the distance to be laid off will be simply the distance from the vertical projection of the point to the ground line.

In like manner, its revolved position will be found when, in the second case, the axis passes through the vertical projection of the point.

REVOLUTION OF THE VERTICAL PLANE.

18. In order to represent both projections of an object on the same sheet of paper or plane, after the projections are made as in the preceding articles, the vertical plane is revolved about the ground line as an axis until it coincides with the horizontal plane, that portion of it which is above the ground line falling beyond it, in the position ABC'D', Fig. 1, and that part which is below coming up in front, in the position ABF'E'.

In this new position of the planes it will be observed, that the planes being regarded as indefinite in extent, all that part of the plane of the paper which is in front of the ground line will represent not only that part of the horizontal plane which is in front of the ground line, but also that part of the vertical plane which is below the horizontal plane; while the part beyond the ground line represents that part of the horizontal plane which is behind the vertical plane, and also that part of the vertical plane which is above the horizontal plane.

19. After the vertical plane is revolved as in the preceding article, the point m', in Fig. 1, will take the position m'' in the line mo produced, and the two projections m and m' will then be in the same straight line, perpendicular to AB. Hence, in every drawing thus made, the *two projections of the same point must be in the same straight line, perpendicular to the ground line.*

If, then, AB, Fig. 7, be the ground line, and it be required to represent or *assume a point in space*, we first assume m for its horizontal projection; through m erect a perpendicular to AB, and assume any point, as m' on this perpendicular, for the vertical projection. The point will then be fully determined, Art. (8). The point thus assumed is in *the first angle*, above the horizontal plane at a distance equal to $m'o$, and in front of the vertical plane at a distance equal to mo, Art. (7).

If the point be in *the second angle*, its horizontal projection m must be on that part of the horizontal plane beyond the ground line, and its vertical projection m' on that part of the vertical plane above the ground line, Art. (5). When the latter plane is revolved to its proper position, m' will fall into that part of the horizontal plane beyond the ground line, and the two projections will be as in (2), Fig. 7, mo representing the distance of the point behind the vertical plane, and $m'o$ its distance above the horizontal plane.

If the point be in *the third angle*, its horizontal projection will be beyond the ground line, and its vertical projection on the part of the vertical plane below the horizontal plane. The vertical plane being revolved to its proper position, m' comes in front of AB, and the two projections will be as in (3).

If the point be in *the fourth angle*, the two projections will be as in (4).

20. To represent, or *assume a plane in space*, we draw at pleasure any straight line, as tT, Fig. 8, to represent its horizontal trace; then through T, draw any other straight line, as Tt', to represent its vertical trace. It is absolutely necessary that these traces intersect AB at the same point, if either intersects it, Art. (9).

The plane and traces being indefinite in extent, the portion included in the first angle is represented by tTt'. The portion in the second angle by $t'Tt''$. That in the third by $t''Tt'''$. That in the fourth by $t'''Tt$.

If the plane be parallel to the ground line and not parallel to either plane of projection, both traces must be assumed parallel to AB, as in Fig. 9; Tt being the horizontal, and Tt' the vertical trace, Art. (9).

If the plane be parallel to either plane of projection, the trace on the other plane is alone assumed, and that parallel to AB.

If the plane be perpendicular to either plane of projection, its trace on this plane may be assumed at pleasure, while its trace on the other plane must be perpendicular to AB, as in (2), Fig. 9; Tt is the horizontal, and Tt' the vertical trace of a plane perpendicular to the horizontal plane.

Also in (3), Tt is the horizontal and Tt' the vertical trace of a plane perpendicular to the vertical plane, Art. (10).

21. To represent or *assume a straight line*, both projections may be drawn at pleasure, as in (1), Fig. 10, mn is the horizontal, and $m'n'$ the vertical projection of a portion of a straight line in *the first angle.*

In (2), mn is the horizontal and $m'n'$ the vertical projection of a portion of a straight line in the *second angle.* In (3), the line is represented in the *third angle*, and in (4), in the *fourth angle.*

In Fig. 11 are, (1), the projections of a right line parallel to the horizontal plane; (2), those of a right line parallel to the vertical plane, (3), those of a right line perpendicular to the hori zontal plane; and (4), those of a right line perpendicular to the vertical plane, Art. (14.)

22. *To assume a point upon a given right line;* since the pro ections of the point must be on the projections of the line, Art (13), and in the same perpendicular to the ground line, Art. (19); we assume the horizontal projection as m, Fig. 17, on mn, and at this point erect mm' perpendicular to AB: m', where it intersects $m'n'$, will be the vertical projection of the point.

23. If two lines intersect, their projections will intersect; for the point of intersection being on each of the lines, its horizontal projection must be on the horizontal projection of each of the lines, Art. (13), and hence at their intersection. For the same reason, the vertical projection of the point must be at the intersection of their vertical projections. These two points being the projection of the same point, must be in the same straight line perpendicular to the ground line, Art. (19). Hence if any two lines intersect in space, *the right line joining the points in which their projections intersect, must be perpendicular to the ground line.*

Therefore, *to assume two right lines which intersect,* we draw at pleasure both projections of the first line, and the horizontal projection of the second line intersecting that of the first; through this point of intersection erect a perpendicular to the ground line until it intersects the vertical projection of this line; through this point draw at pleasure the vertical projection of the second line. Thus, in Fig. 16, assume mn and $m'n'$, also mo; through m erect mm' perpendicular to AB, and through m' draw $m'o'$ at pleasure.

Two parallel right lines are assumed by drawing their pro jections respectively parallel, Art. (16).

NOTATION TO BE USED IN THE DESCRIPTION OF DRAWINGS.

24. Points represented as in Fig. 7, will be described as the point (mm'), the letter at the horizontal projection being always written and read first, or simply as the point M.

Planes given by their traces, as in Fig. 8, will be described as the plane tTt', the middle letter being the one at the intersection of the two traces, and the other letter of the horizontal trace being the first in order. If the traces are parallel to the ground line, or do not intersect it within the limits of the drawing, the same notation will be used, the middle letter being placed on both traces; in the first case, at the left-hand extremity, and in the second case, at the extremity nearest the ground line.

Lines given by their projections, as in Fig. 10, will be described as the line $(mn, m'n')$, the letters on the horizontal projection being first in order; or simply the line MN.

The planes of projection will often be described by the capitals H and V; H denoting the horizontal, and V the vertical plane.

The ground line will be in general denoted by AB.

MANNER OF DELINEATING THE DIFFERENT LINES USED IN THE REPRESENTATION OF MAGNITUDES, OR IN THE CONSTRUCTION OF PROBLEMS.

25. The projections of the same point will be connected by a dotted line, thus ....

Traces of planes which are given or required, when they can be seen from the point of sight,—that is, when the view is not obstructed, either by the planes of projection or by some intervening opaque object,—are drawn full. When not seen, o when they are the traces of auxiliary planes, not the project ing planes of right lines, they will be drawn broken and dotted, thus:

——··——··——··——··——··——

Lines, or portions of lines, either given or required, when seen, will have their projections full. When not seen, or auxiliary, these projections will be broken, thus:

— — — — — — — — — —

In the construction of problems, planes or surfaces which are required will be regarded as transparent, not concealing other parts previously drawn.

All lines or surfaces are regarded as indefinite in extent, unless limited by their form, or a definite portion is considered for a special purpose. Thus the ground line and projections of lines in Fig. 10 are supposed to be produced indefinitely, the lines delineated simply indicating the directions.

CONSTRUCTION OF ELEMENTARY PROBLEMS RELATING TO THE RIGHT LINE AND PLANE.

26. Having explained the manner of representing with accuracy, points, planes, and right lines, we are now prepared to represent the solution of a number of important problems, reating to these magnitudes in space.

In every problem certain points and magnitudes are given, from which certain other points or magnitudes are to be constructed.

Let a right line be first drawn on the paper or slate to represent the ground line; then assume, as in Art. (19), &c., the representatives of the given objects. The proper solution of the problem will now consist of *two distinct parts.* The first is a clear statement of the principles and reasoning to be employed in the construction of the drawing. This is *the analysis* of the problem. The second is the construction, in proper order, of the different lines which are used and required in the problem. This is *the construction* of the problem.

27. PROBLEM 1. *To find the points in which a given right line pierces the planes of projection.*

Let AB, Fig. 12, be the ground line, and (mn, $m'n'$), or simply MN, the given line.

First. To find the point in which this line pierces the horizontal plane.

Analysis. Since the required point is in the horizontal plane, its vertical projection is in the ground line, Art. (8); and since the point is in the given line, its vertical projection will be in the vertical projection of this line, Art. (13); hence it must be at the intersection of this vertical projection with the ground line. The horizontal projection of the required point must be in a straight line drawn through its vertical projection, perpendicular to the ground line, Art. (19), and also in the horizontal projection of the given line; hence it will be at the intersection of these two lines. But the point being in the horizontal plane, is the same as its horizontal projection, Art. (8); hence the rule: *Produce the vertical projection of the line until it intersects the ground line; at the point of intersection erect a perpendicular to the ground line, and produce it until it intersects the horizontal projection of the line;* this point of intersection is the required point.

Construction. Produce $m'n'$ to m'; at m' erect the perpendicular $m'm$, and produce it to m. This is the required point.

Second. In the above analysis, by changing the word "vertical" into "horizontal," and the reverse, we have the analysis and rule for finding the point in which the given line pierces the vertical plane.

Construction. Produce mn to o; at o erect the perpendicular oo', and produce it to o'. This is the required point.

28. Problem 2. *To find the length of a right line joining two given points in space.*

Let AB, Fig. 13, be the ground line, and (mm') and (nn') the two given points.

Analysis. Since the required line contains the two points, its projection must contain the projections of the points, Art. (13). Hence, if we join the horizontal projections of the points by a right line, it will be *the horizontal projection* of the line; and if we join the vertical projections of the points, we shall have *its vertical projection.*

If we now revolve the horizontal projecting plane of the line about its horizontal trace until it coincides with the horizontal plane, and find the revolved position of the points, and join them by a right line, it will be the required distance, since the points do not change their relative position during the revolution, Art. (17).

Construction. Draw mn and $m'n'$. MN will be the required line.

Now revolve its horizontal projecting plane about mn until it coincides with H; the points M and N will fall at m'' and n'', at distances from m and n equal to rm' and sn' respectively, Art. (17); join m'' and n''; $m''n''$ will be the required distance.

Since the point o, in which the line produced pierces H, is in the axis, it remains fixed. The line $m''n''$ produced must then pass through o, and the accuracy of the drawing may thus be verified.

29. *Second method* for the same problem.

Analysis. If we revolve the horizontal projecting plane of the line about the projecting perpendicular of either of its points until it becomes parallel to the vertical plane, the line will, in its revolved position, be projected on this plane in its true length, Art. (14). If we then construct this vertical projection, it will be the required distance.

Construction. Revolve the projecting plane about the perpendicular at m. The point n describes the arc nl, until it comes into the line ml parallel to AB; l will be the horizontal projection of N in its revolved position. Its vertical projection must be in ll' perpendicular to AB; and since, during the revolution, the point N remains at the same distance above H, its vertical projection must also be in the line $n'l'$ parallel to AB, Art. (7), therefore it will be at l'.

The point M being in the axis remains fixed, and its vertical projection remains at m'; $m'l'$ is then the vertical projection of MN in its revolved position, and the true distance.

By examining the drawing, it will be seen that the true distance is the hypothenuse of a right-angled triangle whose base is the horizontal projection of the line, and altitude the difference between the distances of its two extremities from the horizontal plane. Also, that the angle at the base is equal to the angle made by the line with its projection, or the angle made by the line with the horizontal plane. Also, that the length of the line is always greater than that of its projection, unless it is parallel to the plane of projection.

30. Every right line of a plane must pierce any other plane, to which it is not parallel, in the common intersection of the two; hence every right line of a plane, not parallel to the horizontal plane of projection, will pierce it in the horizontal trace of the plane and if not parallel to the vertical plane, will pierce it in the vertical trace: hence, *to assume a straight line in a given plane*, take a point in each trace, and join the two by a right line; or otherwise, draw the horizontal projection at pleasure; at the

points where it intersects the ground line and the horizontal trace erect perpendiculars to the ground line; join the point where the first intersects the vertical trace with the point where the second intersects the ground line: this will be the vertical projection of the line.

Thus, in Fig. 14, draw mn, also mm' and nn': join $m'n'$; it will be the required vertical projection.

31. Problem 3. *To pass a plane through three given points.*

Let M, N, and P, Fig. 15, be the three points.

Analysis. If we join either two of the points by a right line, it will lie in the required plane, and pierce the planes of projection in the traces of this plane, Art. (30). If we join one of these points with the third point, we shall have a second line of the plane. If we find the points in which these lines pierce the planes of projection, we shall have two points of each trace. The traces, and therefore the plane, will be fully determined.

Construction. Join m and n by the straight line mn; also m' and n' by $m'n'$. MN will be the line joining the first two points. This pierces H at h, and V at v, as in Problem 1. Draw also np and $n'p'$; NP will be the second line. It pierces H at t and V at t'. Join h and t, by the straight line ht; it is the required horizontal trace. Join v and t'; $t'v$ is the vertical trace: Or produce ht until it meets AB, and join this point with either v or t' for the vertical trace, Art. (9).

If either MN or NP should be parallel to AB, the plane, and consequently its traces, will be parallel to AB, Art. (9), and it will be necessary to find only one point in each trace.

32. If it be required to pass a plane through two right lines which either intersect or are parallel, we have simply to find the points in which these lines pierce the planes of projection, as in the preceding problem. If the lines do not pierce the planes of

projection, within the limits of the drawing, then any two points of the lines may be joined by a right line, and a point in each trace be determined, by finding the points in which this line pierces the planes of projection.

33. A plane may be passed through a point and right line, by joining the point with any point of the line by a right line, and then passing a plane through these lines; or by drawing through the point a line parallel to the given line, and then passing a plane through the parallels, as above.

34. Problem 4. *To find the angle between two right lines which intersect.*

Let MN and MO, Fig. 16, be the two right lines, assumed as in Art. (23).

Analysis. Since the lines intersect, pass a plane through them, and revolve this plane about its horizontal trace until it coincides with the horizontal plane, and find the revolved position of the two lines. Since they do not change their relative position, their angle, in this new position, will be the required angle.

Construction. The line MN pierces H at *n*, and the line MO at *o*, Art. (27); *no* is then the horizontal trace of the plane containing the two lines, Art. (32). Revolve this plane about *no* until it coincides with H. The point M falls at *p*, Art. (17). The points *n* and *o*, being in the axis, remain fixed; *np* will then be the revolved position of MN, and *po* of NO, and the angle *npo* will be the required angle.

35. *Second method* for the same problem.

Analysis. Revolve the plane of the two lines about its horizontal trace until it becomes perpendicular to the horizontal plane; then revolve it about its new vertical trace until it coincides with the vertical plane; the angle will then be in the vertical plane in its true size.

Construction. First revolve the plane about *no* until it becomes perpendicular to H. T*t'* will be the new vertical trace, the point M will be horizontally projected at *s*, and vertically at *s'*, *s'r* being equal to *sp*.

Now revolve the plane about T*t'* until it coincides with V; *o* will revolve to *o''*, *s* to *s''*, and *n* to *n''*, while the point M, or (*ss'*), will be found at *w*; *n''w* and *o''w* will be the revolved positions of the two lines, and *n''wo''* the required angle.

By examining the drawing, it will be seen that if the angle is oblique, it is less than its projection, unless both lines are parallel to the plane of projection, in which case the angle is equal to its projection.

Let the problem be constructed with one of the lines parallel to the ground line.

36. *If two right lines be perpendicular to each other in space, and one of them parallel to the plane of projection, their projections will be perpendicular.* For the projecting plane of the line which is not parallel to the plane of projection is perpendicular to the second line, and also to its projection, since this projection is parallel to the line itself, Art. (14); and since this projection is perpendicular to this projecting plane, it is perpendicular to its trace, which is the projection of the first line.

37. Problem 5. *To find the position of a line bisecting the angle formed by two right lines, one of which is perpendicular to either plane of projection.*

Let MN and OP, Fig. 17, be the two lines, the latter being perpendicular to the vertical plane.

Analysis. If the plane of the two lines be revolved about the second, until it becomes parallel to the horizontal plane, the angle will be projected on this plane in its true size, and may be bisected by a right line. If the plane be then revolved to its primitive position, and the true position of one point of the bisecting line be determined, and joined with the vertex of the given angle, we shall have the required line.

Construction. Let the plane of the two lines be revolved about OP, until it becomes parallel to H. Any point of MN, as M, will describe the arc of a circle parallel to V, and be horizontally projected at m'', and om'' will be the projection of MN, and $m''op$ will be the true size of the angle. Bisect it by oq, which will be the horizontal projection of the bisecting line in its revolved position. Join m'' with any point of op, as p; this will be the horizontal projection of a line of the given plane, in its revolved position, which intersects the bisecting line in a point horizontally projected at q. When the plane resumes its primitive position, this line will be horizontally projected in mp, and the point, of which q is the horizontal projection, will be horizontally projected at r, and vertically at r'; hence or will be the horizontal, and $o'm'$ the vertical projection of the required line.

Or the plane of the two lines may be revolved about its vertical trace, and the true position determined as indicated in the figure.

38. Problem 6. *To find the intersection of two planes*

Let tTt' and sSs', Fig. 18, be the two planes.

Analysis. Since the line of intersection is a right line, contained in each plane, it must pierce the horizontal plane in the horizontal trace of each plane, Art. (30); that is, *at the intersection of the two traces.* For the same reason, it must pierce the vertical plane at the intersection of the vertical traces. If these two points be joined by a right line, it will be the required intersection.

Construction. The required line pierces H at o and V at p': o is its own horizontal projection, and p' is horizontally projected at p; hence po is the horizontal projection of the required line; o is vertically projected at o'; p' is its own vertical projection; and $o'p'$ is the vertical projection of the required line.

39. *Second Method* for the same problem. When either the horizonta. or vertical traces do not intersect within the limits of the drawing. Let tTt' and sSs', Fig 19, be the planes; tT and sS not intersecting within the limits of the drawing.

Analysis. If we pass any plane parallel to the vertical plane, it will intersect each of the given planes in a line parallel to its vertical trace, and these two lines will intersect in a point of the required intersection. A second point may be determined in the same way, and the right line joining these two points will be the required line.

Construction. Draw pq parallel to AB; it will be the trace of an auxiliary plane. It intersects the two given planes in lines, which pierce H at p and q, and are vertically projected in $p'o'$ and $q'o'$; o' is the vertical, and o the horizontal projection of their intersection. Draw mn also parallel to AB, and thus determine L. OL is the required line. Let the problem be constructed when both planes are parallel to the ground line.

40. Problem 7. *To find the point in which a given right line pierces a given plane.*

Let MN, Fig. 20, be the given line, and tTt' the given plane.

Analysis. If through the line any plane be passed, it will intersect the given plane in a right line, which must contain the required point, Art. (30) This point must also be on the given line; hence it will be at the intersection of the two lines.

Construction. Let the auxiliary plane be the horizontal projecting plane of the line; np is its horizontal and pt' its vertical trace, Art. (10). It intersects tTt' in a right line, which pierces H at o and V at t', of which $o't'$ is the vertical projection, Art. (38). The point m', in which $o't'$ intersects $m'n'$, is the vertical projection of the required point; and m is its horizontal projection. The accuracy of the drawing may be verified by using the vertical projecting plane of MN, as an auxiliary plane, and determining m directly as represented in the figure.

Let the problem be constructed when the given line is parallel to the ground line.

41. *Second Method* for the same problem. When the plane is given by any two of its right lines, find the points in which these two lines pierce either projecting plane of the given line, and join these points by a straight line; this will intersect the given line in the required point.

Construction. Let MN and OP, Fig. 21, be the lines of the given plane, intersecting at L, and QR be the given line. The line MN pierces the horizontal projecting plane of QR at a point of which m is the horizontal, and m' the vertical projection. OP pierces the same plane at P, and $p'm'$ is the vertical projection of the line joining these two points. This intersects $q'r'$ at r', which is the vertical projection of the required point, and r its horizontal projection.

42. If either projection of a point of an oblique plane be given, the other projection may at once be determined by a simple application of the principles of the preceding problem. Thus let m, Fig. 22, be the horizontal projection of a point of the plane tTt'. If at m a perpendicular be erected to H, it will pierce tTt' in the only point of the plane which can be horizontally projected at m; m is the horizontal, and $m''m'$ the vertical projection of this perpendicular. Through it pass any plane, as that whose horizontal trace is no. Since this plane is perpendicular to H, nn' will be its vertical trace. It intersects tTt' in a right line, of which $o'n'$ is the vertical projection; hence m' is the required vertical projection, Art. (40.)

The auxiliary plane may be passed parallel to tT; mp will be its horizontal, and pp' its vertical trace. It intersects tTt' in a line parallel to tT, which pierces V at p', of which mp is the horizontal, and $m'p'$ the vertical projection, and m' will be the required vertical projection.

In a similar way, if the vertical projection be given, the horizontal can be found.

43. *If a right line is perpendicular to a plane, its projections will be respectively perpendicular to the traces of the plane.*

For the horizontal projecting plane of the line is perpendicular to the given plane, since it contains a line perpendicular to it. This projecting plane is also perpendicular to the horizontal plane, Art. (11). It is therefore perpendicular to the intersection of these two planes, which is the horizontal trace of the given plane. Hence *the horizontal projection of the line*, which is a line of this projecting plane, must be perpendicular *to the horizontal trace.*

In the same way it may be proved that *the vertical projection* of the line will be perpendicular *to the vertical trace.*

Conversely, *if the projections of a right line are respectively perpendicular to the traces of a plane, the line will be perpendicular to the plane.*

For, if through the horizontal projection of the line, its horizontal projecting plane be passed, it will be perpendicular to the horizontal trace of the given plane, and therefore perpendicular to the plane. In the same way it may be proved that the vertical projecting plane of the line is perpendicular to the given plane; therefore the intersection of these two planes, which is *the given line*, is perpendicular to the given plane.

Hence, *to assume a right line perpendicular to a plane*, we draw its projections perpendicular to the traces of the plane respectively.

Also, *to assume a plane perpendicular to a right line*, we draw the two traces from any point in the ground line, perpendicular to the projections of the line.

44. PROBLEM 8. *To draw through a given point a right line perpendicular to a given plane, and to find the distance of the point from the plane.*

Let M, Fig. 23, be the given point, and tTt' the plane.

Analysis. Since the required perpendicular is to pass through the given point, its projections must pass through the projections of the point, Art. (13); and since it is to be perpendicular to the plane, these projections must be respectively perpendicular to the traces of the plane, Art. (43). Hence, if through the horizontal projection of the point, a right line be drawn perpendicular to the horizontal trace, and through the vertical projection, a right line perpendicular to the vertical trace, they will be respectively the horizontal and vertical projections of the required line.

If the point in which this perpendicular pierces the plane be found, the distance between this point and the given point will be the required distance or length of the perpendicular.

Construction. Through m draw mn perpendicular to $t'T$, and through m', $m'n'$, perpendicular to $t''T$. MN will be the required perpendicular. N is the point in which MN pierces the plane, Art. (40), and $m''n''$ the length of the perpendicular, Art. (28).

Let the problem be constructed when the plane is parallel to the ground line; also when it is perpendicular to it.

45. PROBLEM 9. *To project a given right line on any oblique plane, and to show the true position of this projection.*

Let MN, Fig. 24, be the given line, and tTt' the given plane.

Analysis. If through any two points of the line perpendiculars be drawn to the plane, and the points in which they pierce the plane be found, these will be two points of the required projection, and the right line joining them will be the required line. If now the plane be revolved about its horizontal trace to coincide with the horizontal plane, or about its vertical trace until it coincides with the vertical plane, and the revolved position of these two points be found and joined by right line, this will show the true position of the line in the oblique plane.

Construction. Assume the two points M and P, Art. (22), and draw the perpendiculars MR and PS, Art. (44). The first pierces the plane at R, and the second at S, Art. (40); and rs will be the horizontal, and $r's'$ the vertical projection of the required projection. The point N, in which the given line pierces the plane, will also be one point of the required projection.

Now revolve the plane about tT until it coincides with H. R is found at r'', Art. (17), and S at s'', and $r''s''$ is the true position of RS in its own plane: $r''s''$ produced must pass through the point in which the projection pierces H.

If the given line be parallel to the plane, it will only be necessary to determine the projection of one point on the plane, and through this to draw a line parallel to the given line, Art. (14).

46. PROBLEM 10. *Through a given point, to pass a plane perpendicular to a given right line.*

Let M, Fig. 25, be the given point, and NO the given line.

Analysis. Since the plane is to be perpendicular to the line, its traces must be respectively perpendicular to the projections of the line, Art. (43). We thus know the direction of the traces. Through the point, draw a line parallel to the horizontal trace; it will be a line of the required plane, and will pierce the vertical plane in a point of the vertical trace. Through this point draw a right line perpendicular to the vertical projection of the line; it will be the vertical trace of the required plane. Through the point in which this trace intersects the ground line, draw a right line perpendicular to the horizontal projection of the line; it will be the horizontal trace.

Construction. Through m draw mp, perpendicular to no; it will be the horizontal projection of a line through M, parallel to the horizontal trace; and since this line is parallel to H, its vertical projection will be $m'p'$, parallel to AB. This line pierces V at p', Art. (27). Draw p'T perpendicular to $n'o'$, and Tt perpendicular to no; tTp' will be the required plane. Or, through M draw MS parallel to the vertical trace. It pierces H at s, which must be a point of the horizontal trace, and the accuracy of the drawing may thus be tested.

47. PROBLEM 11. *To pass a plane through a given point, parallel to two given right lines.*

Let M, Fig. 26, be the point, and NO, and PQ, the two given lines.

Analysis. Through the given point, draw a line parallel to each of the given lines. The plane of these two lines will be the required plane, since it contains a line parallel to each of the given lines.

Construction. Through m draw ms, parallel to no, and through m', $m's'$ parallel to $n'o'$. The line MS will be parallel to NO, Art (16). In the same way, construct MR parallel to QP. These lines pierce H at s and t respectively, and MR pierces V at r'; hence tTr' is the required plane, Art. (32).

Let the problem be constructed when one of the given lines is parallel to the ground line.

Let the problem, to pass a plane through a given point parallel to a given plane, also be constructed.

48. PROBLEM 12. *To pass a plane through a given right line, parallel to another right line.*

Let MN, Fig. 27, be the line through which the plane is to be passed, and PQ the other given line.

Analysis. Through any point of the first line, draw a line parallel to the second. Through this auxiliary line and the first, pass a plane. It will be the required plane.

Construction. Through R, on the first line, draw RO parallel to PQ It pierces H at o, and V at t'. MN pierces H at m, and V at n'; hence oTt' is the required plane.

Let the problem be constructed when either line is parallel to the ground line.

49. PROBLEM 13. *To find the shortest distance from a given point to a given right line.*

Let M, Fig. 28, be the given point, and NO the given straight line.

Analysis. The required distance is the length of a perpendicular from the point to the line. If through the given point and the line we pass a plane, and revolve this plane about either trace until it coincides with the corresponding plane of projection, the line and point will not change their relative positions; hence, if through the revolved position of the point we draw a perpendicular to the revolved position of the line, it will be the required distance.

Construction. Through M draw MP parallel to NO. It pierces H at p. NO pierces H at o. po is then the horizontal trace of the plane through M and NO, Art. (32). Revolve this plane about op until it coincides with H. M falls at m'', Art. (17). Since p remains fixed, pm'' is the revolved position of MP NO being parallel to MP before revolution, will be parallel after; and as o is in the axis, oq'' parallel to pm'' will be the revolved position of NO. Draw $m''q''$ perpendicular to oq''; *it will be the required distance.* When the plane is revolved back to its primitive position, m'' is horizontally projected at m, and q'' at q hence MQ is the perpendicular in its true position.

50. *Second method* for the same problem.

Analysis. If through the given point a plane be passed perpendicular to the given line, Art. (46), and the point in which the given line pierces the plane be found, Art. (40), and joined with the given point, we shall have the required distance, the true length of which can be found as in Art. (28).

Let the problem be constructed in accordance with this analysis.

Let the problem also be constructed when the given line is parallel to the horizontal plane.

51. PROBLEM 14. *To find the angle which a given right line makes with a given plane.*

Let MN, Fig. 29, be the given line, and tTt' the given plane.

Analysis. The angle made by the line with the plane, is the same as that made by the line with its projection on the plane. Hence, if through any point of the line a perpendicular be drawn to the plane, the foot of this perpendicular will be one point of the projection. If this point be joined with the point in which the given line pierces the plane, we shall have the projection of the line on the plane, Art. (45). This projection, the perpendicular, and a portion of the given line, form a right-angled triangle; of which the projection is the base, and the angle at the base is the required angle. But the angle at the vertex, that is, *the angle between the perpendicular and given line*, is the complement of the required angle; hence, if we find the latter angle, and subtract it from a right angle, we shall have the required angle.

Construction. Through M draw the perpendicular MP to tTt' Art. (44). It pierces H in p. The given line pierces H in o, and op is the horizontal trace of the plane of the two lines, Art. (32). Revolve this plane about op, and determine their angle $pm''o$, as in Art. (34). Its complement, prm', is equal to the required angle.

Let the problem be constructed when the plane is parallel to the ground line.

52. PROBLEM 15. *To find the angle between two given planes.*

Let sSs', Fig. 30, and tTt', be the two planes, intersecting in the line ON, Art. (38).

Analysis. If we pass a plane perpendicular to the intersection of the two planes, it will be perpendicular to both; and cut from each a right line perpendicular to this intersection at a common point. The angle between these lines will be the measure of the required angle.

Construction. Draw pq perpendicular to on; it will be the horizontal trace of a plane perpendicular to ON, Art. (43). This plane intersects the given planes in right lines, one of which pierces H at p, and the other at q. If right lines be drawn from these points to the point in which the auxiliary plane intersects ON, they will be the lines cut from the planes, and the angle between them will be the required angle.

The vertical trace of the auxiliary plane may be drawn as in Art. (43), and the vertex of the angle found as in Art. (40), and then the angle as in Art. (34). Or otherwise, thus: Suppose a right line to be drawn from r to the vertex of the angle, it will be perpendicular to ON, since it is contained in a plane perpendicular to it; it will also be perpendicular to pq, since it is in the horizontal projecting plane of ON, which is perpendicular to pq, Art. (43). If this projecting plane be revolved about no until it coincides with H, n' will fall at n''; and since o is fixed, on'' will be the revolved position of ON, and rm'', perpendicular to on'', will be the revolved position of the line joining r with the vertex. If now the plane of the two lines be revolved about pq until it coincides with H, m'' will be at v, rv being equal to rm'', and pvq will be the required angle, Art. (34).

The point m'', from its true position, is horizontally projected at m, and vertically at m', and pmq is the horizontal, and $p'm'q'$ the vertical projection of the angle.

Let the problem be constructed when both planes are parallel to the ground line.

53. If the angle between a given plane and either plane of projection, as the horizontal, be required, we simply pass a plane perpendicular to the horizontal trace, as in Fig. 31. This plane cuts *on* from H, and ON from *t*T*t'*, and the angle *non''*, found by revolving the auxiliary plane about *on*, Art. (34), will be the required angle.

In the same way the angle *p'q'p''*, between the given plane and vertical plane, may be found.

54. Problem 16. *Either trace of a plane being given, and the angle which the plane makes with the corresponding plane of projection, to construct the other trace.*

Let *t*T, Fig. 31, be the horizontal trace of the plane, and *def* the angle which the plane makes with the horizontal plane.

Analysis. If a right line be drawn through any point of the given trace, perpendicular to it, it will be the horizontal trace of a plane perpendicular to the given trace, and if at the same point a line be drawn, making with this line an angle equal to the given angle, this will be the revolved position of a line cut from the required plane by this perpendicular plane, Art. (53). If this line be revolved to its true position, and the point in which it pierces the vertical plane be found, this will be a point of the required vertical trace. If this point be joined with the point where the horizontal trace intersects the ground line, we shall have the vertical trace.

Construction. Through *o* draw *no* perpendicular to *t*T; also *on''*, making the angle *non''* = *def*; *on''* will be the revolved position of a line of the required plane. When this line is revolved to its true position, it pierces V at *n'*, and *n'*T is the required trace.

If the given trace does not intersect the ground line within the limits of the drawing, the same construction may be made at a second point of the trace, and thus another point of the vertical trace be determined.

55. PROBLEM 17. *To find the shortest line which can be drawn, terminating in two right lines, not in the same plane.*

Let MN, Fig. 32, and OP, be the two right lines.

Analysis. The required line is manifestly a right line, perpendicular to both of the given lines. If through one of the lines we pass a plane parallel to the other, and then project this second line on this plane, this projection will be parallel to the line itself, Art. (14), and therefore not parallel to the first line. It will then intersect the first line in a point. If at this point we erect a perpendicular to the plane, it will be contained in the projecting plane of the first line, be perpendicular to both lines, and intersect them both. That portion included between them is the required line.

Construction. Through MN pass a plane parallel to OP, Art. (48): mr is its horizontal, and $k'n'$ its vertical trace. Through any point of OP, as Q, draw QU perpendicular to this plane, Art. (49). It pierces the plane at U, Art. (40); and this is one point of the projection of OP on the parallel plane. Through U draw UX parallel to OP; it will be the projection of OP on the plane. It intersects MN in X, which is the point through which the required line is to be drawn; and XY, perpendicular to the plane, is the required line, the true length of which is $x''y''$, Art. (28).

Let the problem be constructed with one of the lines parallel to the ground line.

Also with one of the lines perpendicular to either plane of projection.

56. *Second construction* of the same problem.

Let MN and OP, Fig. 33, be the right lines.

Through MN pass a plane parallel to OP, Art. (48): *mr* is its horizontal trace. Through *p* conceive a perpendicular to be drawn to this plane. The point in which it pierces the plane will be one point of the projection of OP on the plane. To find this point, through the perpendicular pass a plane perpendicular to OP; *pq* will be its horizontal trace, Art. (43). This plane will intersect the parallel plane in a right line, which pierces H at *q*. It intersects the horizontal projecting plane of OP in a right line perpendicular to OP at *p*. To determine this line, revolve the projecting plane of OP about *op* until it coincides with H. Any point of OP, as L, falls at l'', and pl'' is the revolved position of OP. This projecting plane intersects the parallel plane in a right line, which pierces H at *k*, and is parallel to OP; *ku*, parallel to pl'', is the revolved position of this parallel line; *pu*, perpendicular to pl'', is the revolved position of the intersection of the projecting plane and perpendicular plane; and *u* is the revolved position of a point of the line of intersection of the perpendicular and parallel plane. Now revolve the plane perpendicular to OP about *pq* as an axis, until it coincides with H. The point, of which *u* is the revolved position, falls at u'', and $u''q$ is the revolved position of the line of intersection of the perpendicular and parallel plane; pp'' is the revolved position of the line through *p* perpendicular to the parallel plane, and is equal to the distance required; and p'' is the revolved position of the projection of *p* on the parallel plane. In the counter-revolution, the point p'' will be horizontally projected, somewhere in the perpendicular to the axis *pq*; $p''x''$ is the horizontal projection of the projection of OP on the parallel plane, and *xy*, perpendicular to *mr*, is the horizontal, and $x'y'$ the vertical projection of the required line.

CONSTRUCTION AND CLASSIFICATION OF LINES.

57. *Every line may be generated by the continued motion of a point.* If the generating point be taken in any position on th line, and then be moved to its next position, these two points may be regarded as forming an *infinitely small right line*, or *elementary line.* The two points are *consecutive points*, or points having no distance between them, and may practically be considered as one point.

The line may thus be regarded as made up of an infinite number of infinitely small elements, each element indicating the direction of the motion of the point while generating that part of the line.

58. The law which directs the motion of the generating point, determines the nature and class of the line.

If the point moves always in the same direction, that is, so that *the elements of the line are all in the same direction*, the line generated is *a right line.*

If the point moves so as continually to change its direction from point to point, the line generated is *a curved line* or *curve.*

If *all the elements of a curve are in the same plane*, the curve is *of single curvature.*

If *no three consecutive elements*, that is, if no four consecutive points *are in the same plane*, the curve is *of double curvature.*

We thus have three general classes of lines.

I. Right lines: all of whose points lie in the same direction.

II. Curves of single curvature: all of whose points lie in the same plane.

III. Curves of double curvature: no four consecutive points of which lie in the same plane.

59. The simplest curves of single curvature are:

I. *The circumference of a circle*, which may be generated by a point moving in the same plane, so as to remain at the same distance from a given point.

II. *A parabola*, which may be generated by a point moving in the same plane, so that its distance from a given point shall be constantly equal to its distance from a given right line.

The given point is *the focus*, the given right line *the directrix*.

If through the focus a right line be drawn perpendicular to the directrix, it is *the axis* of the parabola; and the point in which the axis intersects the curve is *the vertex*.

From the definition, the curve may readily be constructed by points, thus: Let F, Fig. 34, be the focus, and CD the directrix. Through F draw FC perpendicular to CD. It will be *the axis*. The point V, midway between F and C, is a point of the curve, and is *the vertex*. Take any point on the axis, as P, and erect the perpendicular PM to the axis. With F as a centre, and CP as a radius, describe an arc cutting PM in the two points M and M'. These will be points of the curve, since

$$FM = CP = DM, \text{ also } FM' = CP = D'M'.$$

In the same way all the points may be constructed.

III. *An ellipse*, which may be generated by a point moving in the same plane, so that the sum of its distances from two fixed points shall be constantly equal to a given right line.

The two fixed points are *the foci*. The curve may be constructed by points, thus: Let F and F', Fig. 35, be the two foci, and VV' the given right line, so placed that VF = V'F'.

Take any point as P between F and F'. With F as a centre, and V'P as a radius, describe an arc. With F' as a centre, and VP as a radius, describe a second arc, intersecting the first in the points M and M'. These will be points of the required curve, since

$$MF + MF' = VP + V'P = VV'; \text{ also } MF + M'F' = VV'.$$

In the same way all the points may be constructed. V and V′ are evidently points of the curve, since

$$VF + VF' = V'F' + VF' = VV'; \text{ also } V'F' + V'F = VV'.$$

The point C, midway between the foci, is *the centre* of the curve. The line VV′, passing through the foci, and terminating in the curve, is *the transverse axis* of the curve.

The points V and V′ are *the vertices* of the curve. DD′ perpendicular to VV′, at the centre, is *the conjugate axis* of the curve.

If the two axes are given, the foci may be constructed thus: With D the extremity of the conjugate axis as a centre, and CV the semi-transverse axis as a radius, describe an arc cutting VV′ in F and F′. These points will be the foci, for

$$DF + DF' = 2CV = VV'.$$

IV. *The hyperbola*, which may be generated by moving a point in the same plane, so that the difference of its distances from two fixed points shall be equal to a given line.

The two fixed points are *the foci*.

The curve may be constructed by points, thus: Let F and F′, Fig. 36, be the two foci, and VV′ the given line, so placed that FV = F′V′.

With F′ as a centre, and any radius greater than F′V, as F′O, describe an arc. With F as a centre, and a radius FM, equal to F′O — VV′, describe a second arc, intersecting the first in the points M and M′. These will be points of the required curve, since

$$F'M - FM = F'O - FM = VV'; \text{ also } F'M' - FM' = VV$$

In the same way any number of points may be determined.

It is manifest, also, that if the greater radius be used with F as a centre, another branch, N'V'N', exactly equal to MVM', will be described.

V and V' are evidently points of the curve, since

$$F'V - FV = VV' = FV' - F'V',$$

and are *the vertices of the hyperbola.*

The point C, midway between the foci, is *the centre*, and VV' is *the transverse axis.* A perpendicular, DD', to the transverse axis at the centre, is *the indefinite conjugate axis.* It evidently does not intersect the curve.

PROJECTION OF CURVES.

60. If all the points of a curve be projected upon the horizontal plane, and these projections be joined by a line, this line is *the horizontal projection of the curve.*

Likewise, if the vertical projections of all the points of a curve be joined by a line, it will be *the vertical projection of the curve.*

61. *The two projections of a curve being given, the curve will, in general, be completely determined.* For in the same perpendicular to the ground line, two points, one on each projection, may be assumed, and the corresponding point of the curve determined, as in Art. (8). Thus *m* and *m'*, Fig. 37, being assumed in a perpendicular to AB, M will be a point of the curve, and in the same way every point of the curve may in general be determined.

62. If the plane of a curve of single curvature is perpendicular to either plane of projection, the projection of the curve on that plane will be a right line, and all of its points will be projected into the trace of the plane on this plane of projection.

If the plane of the curve be perpendicular to the ground line, both projections will be right lines, perpendicular to the ground line, and the curve will be undetermined, as in Art. (15).

If the plane of the curve be parallel to either plane of projection, its projection on that plane will be equal to itself, since each element of the curve will be projected into an equal element, Art. (14). Its projection on the other plane will be a right line, parallel to the ground line.

The projection of a curve of double curvature can in no case be a right line.

63. The points in which a curve pierces either plane of projection can be found by the same rule as in Art. (27). Thus o, Fig. 37, is the point in which the curve MN pierces H, and p' the point in which it pierces V.

TANGENTS AND NORMALS TO LINES.

64. If a right line be drawn through any point of a curve, as M, Fig. 38, intersecting it in another point, as M′, and then the second point be moved along the curve towards M, until it coincides with it, the line, during the motion containing both points, will become *tangent to the curve* at M, which is *the point of contact.*

As when the point M′ becomes consecutive with M, the line thus containing the element of the curve at M, Art. (57), may, for all practical purposes, be regarded as the tangent, we say that *a right line is tangent to another line, when it contains two consecutive points of that line.*

If a right line continually approaches a curve, and becomes tangent to it, at an infinite distance, it is called *an asymptote of the curve.*

Two curves are tangent to each other, when they contain two consecutive points, *or have, at a common point, a common tangent.*

If a right line is tangent to a curve of single curvature, it will be contained in the plane of the curve, for it passes through two points in that plane, viz., the two consecutive points of the curve.

Also, *if a right line is tangent to another right line, it will coincide with it*, as the two lines have two points in common.

The expression, "a tangent to a curve," or "a tangent," will hereafter be understood to mean a rectilineal tangent, unless otherwise mentioned.

65. *If two lines are tangent in space, their projections on the same plane will be tangent to each other.* For the projections of the two consecutive points, common to the two lines, will also be consecutive points, common to the projections of both lines, Art. (60).

The converse of this is not necessarily true. But if both the horizontal and vertical projections are tangent at points, which are the projections of a common point of the two lines, Art. (23), the lines will be tangent in space; for the projecting perpendiculars, at the common consecutive points, will intersect in two consecutive points common to the two lines.

66. If a right line be drawn perpendicular to a tangent at its point of contact, as MO, Fig. 38, it is *a normal to the curve.* As an infinite number of perpendiculars can be thus drawn, all in a plane perpendicular to MT at M, there will be an infinite number of normals at the same point.

If the curve be a plane curve, that is, a curve of single curvature, the term "normal" will be understood to mean that normal which is in the plane of the curve, unless otherwise mentioned.

67. If we conceive a curve to be rolled on its tangent at any point, until each of its elements in succession comes into this tangent, the curve is said to be *rectified;* that is a right line, equal to it in length, has been found.

Since the tangent to a curve at a point contains the element of the curve, the angle which the curve, at this point, makes with any line or plane will be the same as that made by the tangent.

THE HELIX.

68. If a point be moved uniformly around a right line, remaining always at the same distance from it, and having at the same time a uniform motion in the direction of the line, it will generate a curve of double curvature, called *a helix.*

The right line is *the axis* of the curve.

Since all the points of the curve are equally distant from the axis, the projection of the curve on a plane perpendicular to this axis will be the circumference of a circle.

Thus let m, Fig. 39, be the horizontal, and $m'n'$ the vertical projection of the axis, and P the generating point, and suppose that while the point moves once around the axis, it moves through the vertical distance $m'n'$; $prqs$ will be the horizontal projection of the curve.

To determine the vertical projection, divide $prqs$ into any number of equal parts, as 16, and also the line $m'n'$ into the same number, as in the figure. Through these points of division draw lines parallel to AB. Since the motion of the point is uniform, while it moves one-eighth of the way round the axis it will ascend one-eighth of the distance $m'n'$, and be horizontally projected at x, and vertically at x'. When the point is horizontally projected at r, it will be vertically projected at r'; and in the same way the points y', q', &c., may be determined, and $p'r'q's'$ will be the required vertical projection.

69. It is evident from the nature of the motion of the generating point, that in generating any two equal portions of the curve, it ascends the same vertical distance; that is, any two elementary

arcs of the curve will make equal angles with the horizontal plane. Thus, if CD (a), Fig. 39, be any element of the curve, the angle which it makes with the horizontal plane will be DCe, or the angle at the base of a right-angled triangle of which Ce = cd, Fig. 39, is the base and De the altitude. But from the nature of the motion, Ce is to De as any arc px is to the corresponding ascent $x''x'$. Hence, if we rectify the arc xp, Art. (67), and with this as a base construct a right-angled triangle, having $x'x''$ for its altitude, the angle at the base will be the angle which the arc, or its tangent at any point, makes with the horizontal plane. Therefore, to draw a tangent at any point as X, we draw xz tangent to the circle pxr at x; it will be the horizontal projection of the required tangent. On this, from x, lay off the rectified arc xp to z; z will be the point where the tangent pierces H, and $z'x'$ will be its vertical projection.

Since the angle which a tangent to the helix makes with the horizontal plane is constant, and since each element of the curve is equal to the hypothenuse of a right-angled triangle of which the base is its horizontal projection, the angle at the base, the constant angle, and the altitude, the ascent of the point while generating the element; it follows, that when the helix is rolled out on its tangent, the sum of the elements, or length of any portion of the curve, will be equal to the hypothenuse of a right-angled triangle, of which the base is its horizontal projection rectified, and altitude, the ascent of the generating point while generating the portion considered. Thus the length of the arc pX is equal to the length of the portion of the tangent ZX.

GENERATION AND CLASSIFICATION OF SURFACES.

70. A surface *may be generated by the continued motion of a line.* The moving line is *the generatrix* of the surface; and the different positions of the generatrix are *the elements.*

If the generatrix be taken in any position, and then be moved to its next position on the surface, these two positions are *consecutive positions* of the generatrix, or *consecutive elements* of the surface, and may practically be regarded as one element.

71. The form of the generatrix, and the law which directs its motion, determine the nature and class of the surface.

Surfaces may be divided into *two* general classes.

First. *Those which can be generated by right lines; or which have rectilinear elements.*

Second. *Those which can only be generated by curves, and which can have no rectilinear elements.* These are DOUBLE CURVED SURFACES.

Those which can be generated by right lines are:

First. PLANES, which may be generated by *a right line moving so as to touch another right line, having all its positions parallel to its first position.*

Second. SINGLE CURVED SURFACES, which may be generated by *a right line, moving so that any two of its consecutive positions shall be in the same plane.*

Third. WARPED SURFACES, which may be generated by *a right line moving so that no two of its consecutive positions shall be in the same plane.*

72. *Single curved surfaces are of three kinds.*

I. Those in which *all the positions* of the rectilinear generatrix are parallel.

II. Those in which *all the positions* of the rectilinear generatrix intersect in a common point.

III. Those in which the consecutive positions of the rectilinear generatrix intersect *two and two*, no three positions intersecting in a common point.

CYLINDRICAL SURFACES, OR CYLINDERS.

73. Single curved surfaces of the first kind are *Cylindrical surfaces, or Cylinders.* Every cylinder may be generated by moving *a right line so as to touch a curve, and have all its positions parallel.*

The moving line is *the rectilinear generatrix.* The curve is *the directrix.* The different positions of the generatrix are *the rectilinear elements of the surface.*

Thus, Fig. 40, if the right line MN be moved along the curve *mlo*, having all its positions parallel to its first position, it will generate a cylinder.

If the cylinder be intersected by any plane not parallel to the rectilinear elements, the curve of intersection may be taken as *a directrix*, and any rectilinear element as the generatrix, and the surface be re-generated. This curve of intersection may also be *the base of the cylinder.*

The intersection of the cylinder by the horizontal plane is usually taken as the base. If this base have a centre, the right line through it, parallel to the rectilinear elements, is *the axis of the cylinder.*

A definite portion of the surface included by two parallel planes is sometimes considered; in which case the lower curve of intersection is *the lower base*, and the other *the upper base.*

Cylinders are distinguished by the name of their bases; as a cylinder with a circular base; a cylinder with an elliptical base.

If the rectilinear elements are perpendicular to the plane of the base, the cylinder is *a right cylinder*, and the base *a right section.*

A cylinder may also be generated by moving the curvilinear directrix, as a generatrix, along any one of the rectilinear elements, as a directrix, the curve remaining always parallel to its first position.

If the curvilinear directrix be changed to a right line, the cylinder becomes a plane.

It is manifest that if a plane parallel to the rectilinear elements intersects the cylinder, the lines of intersection will be rectilinear elements, which will intersect the base.

74. It will be seen that the projecting lines of the different points of a curve, Art. (60), form a right cylinder, the base of which, in the plane of projection, is the projection of the curve.

These cylinders are respectively *the horizontal* and *vertical projecting cylinders* of the curve, and by their intersection determine the curve.

75. A cylinder is represented by projecting one or more of the curves of its surface, and its principal rectilinear elements.

When these elements are not parallel to the horizontal plane, it is usually represented thus: Draw the base, as *mlo*, Fig. 40, in the horizontal plane. Tangent to this, draw right lines *lx* and *kr*, parallel to the horizontal projection of the generatrix; these will be the horizontal projections of the extreme rectilinear elements, as seen from the point of sight, thus forming the horizontal projection of the cylinder. Draw tangents to the base, perpendicular to the ground line, as *mm'*, *oo'*; through the points *m'* and *o'*, draw lines *m'n'* and *o's'*, parallel to the vertical projection of the generatrix, thus forming the vertical projection of the cylinder; *m'o'* being the vertical projection of the base.

76. *To assume a point of the surface*, we first assume one of its projections, as the horizontal. Through this point, erect a perpendicular to the horizontal plane. It will pierce the surface in the only points which can be horizontally projected at the point taken. Through this perpendicular, pass a plane parallel to the

rectilinear elements; it will intersect the cylinder in elements, Art. (73), which will be intersected by the perpendicular in the required points.

Construction. Let p, Fig. 40, be the horizontal projection assumed. Through p, draw pq parallel to lx; it will be the horizontal trace of the auxiliary plane. This plane intersects the cylinder in two elements; one of which pierces H at q, and the other at u; and $q'y'$, and $u'z''$, will be the vertical projections of these elements, $v'p'$ the vertical projection of the perpendicular, and p' and p'' the vertical projections of the two points of the surface, horizontally projected at p.

To *assume a rectilinear element*, we have simply to draw a line parallel to the rectilinear generatrix, through any point of the base, or of the surface.

CONICAL SURFACES, OR CONES.

77. Single curved surfaces of the second kind, are *Conical surfaces*, or *Cones.*

Every cone may be generated by moving a right line so as continually to touch a given curve, and pass through a given point not in the plane of the curve.

The moving line is *the rectilinear generatrix;* the curve, *the directrix;* the given point, *the vertex* of the cone; and the different positions of the generatrix, *the rectilinear elements.*

The generatrix being indefinite in length, will generate two parts of the surface, on different sides of the vertex, which are called *nappes;* one, *the upper*, the other, *the lower nappe.*

Thus, if the right line MS, Fig. 41, move along the curve *mlo* and continually pass through S, it will generate a cone.

If the cone be intersected by any plane not passing through the vertex, the curve of intersection may be taken as *a directrix*, and any rectilinear element as *a generatrix*, and the cone be regenerated. This curve of intersection may also be *the base* of the cone. The intersection of the cone by the horizontal plane, is usually taken as the base.

If a definite portion of the cone included by two parallel planes is considered, it is called *a frustum of a cone;* one of the limiting curves being *the lower*, and the other *the upper base* of the frustum.

Cones are distinguished by the names of their bases; as a cone with a circular base; a cone with a parabolic base, &c.

If the rectilinear elements all make the same angle with a right line passing through the vertex, the cone is *a right cone*, the right line being *its axis.*

A cone may also be generated by moving a curve so as continually to touch a right line, and change its size according to a proper law.

If the curvilinear directrix of a cone be changed to a right line, or if the vertex be taken in the plane of the curve, the cone will become a plane.

If the vertex be removed to an infinite distance, the cone will evidently become a cylinder.

If a cone be intersected by a plane through the vertex, the lines of intersection will be rectilinear elements, intersecting the base.

78 A cone is represented by projecting the vertex, one of the curves on its surface, and its principal rectilinear elements. Thus, let S, Fig. 41, be the vertex. Draw the base, *mlo*, in the horizontal plane, and tangents to this base through *s*, as *sl* and *sk*; thus forming the horizontal projection of the cone. Draw tangents to the base, perpendicular to the ground line, as *mm′*, *oo′*; and through *m′* and *o′*, draw the right lines *m′s′* and *o′s′*, thus forming the vertical projection of the cone.

79. *To assume a point of the surface*, we first assume one of its projections, as the horizontal. Through this erect a perpen

dicular to the horizontal plane; it will pierce the surface in the only points which can be horizontally projected at the point taken. Through this perpendicular and the vertex pass a plane. It will intersect the cone in elements which will be intersected by he perpendicular in the required points.

Construction. Let p be the horizontal projection. Draw ps. It will be the horizontal trace of the auxiliary plane, which intersects the cone in two elements; one of which pierces H at q, and the other at r, and $q's'$ and $r's'$ are the vertical projections of these elements, and p' and p'' are the vertical projections of the two points of the surface.

To assume a rectilinear element, we have simply to draw through any point of the base, or of the surface, a right line to the vertex.

80. Single curved surfaces of the third kind, may be generated by drawing a system of tangents to any curve of double curvature. These tangents will evidently be rectilinear elements of a single curved surface. For if we conceive a series of consecutive points of a curve of double curvature, as a, b, c, d, &c., the tangent which contains a and b, Art. (64), is intersected by the one which contains b and c, at b; that which contains b and c, by the one which contains c and d, at c; and so on, each tangent intersecting the preceding consecutive one, but not the others, since no two elements of the curve, not consecutive, are, in general, in the same plane, Art. (58).

81. If the curve to which the tangents are drawn, is a helix, Art. (68), the surface may be represented thus: Let pxy, Fig. 42, be the horizontal, and $p'x'y'$ the vertical projection of the helical directrix. Since the rectilinear elements are all tangent to this directrix, any one may be assumed, as in Art. (69); hence XZ, YU, &c., are elements of the surface.

A point of the surface may be assumed by taking a point on any assumed element.

These elements pierce H in the points *z*, *u*, *v*, &c., and *zuvw* is the curve in which the surface is intersected by the horizontal plane, and may be regarded as the base of the surface; and it is evident that if the surface be intersected by any plane parallel to this base, the curve of intersection will be equal to the base.

WARPED SURFACES WITH A PLANE DIRECTER.

82. There is a great variety of warped surfaces, differing from each other in their mode of generation and properties.

The most simple are those which may be generated by a right line *generatrix*, moving so as to touch two other lines *as directrices*, and parallel to a given plane, called *a plane directer.*

Such surfaces are *warped surfaces, with two linear directrices and a plane directer.*

They, as all other warped surfaces, may be represented by projecting one or more curves of the surface, and the principal rectilinear elements.

83. The directrices and plane directer being given, *a rectilinear element, passing through any point of either directrix, may be determined*, by passing a plane through this point, parallel to the plane directer, and finding the point in which this plane cuts the other directrix, and joining this with the given point.

Construction. Let MN and PQ, Fig. 43, be any two linear directrices, *t*T*t'* the plane directer, and O any point of the first directrix.

Assume any line of the plane directer, as CD, Art. (30), and through the different points of this line, draw right lines SE, SF, &c., to any point, as *s* of *t*T. Through O draw a system of lines OR, OY, &c., parallel respectively to SE, SF, &c. These will form a plane through O, parallel to tTt', and pierce the horizontal projecting cylinder of PQ, in the points R, Y, W, &c. These points being joined, will form the curve RW, which intersects PQ in X, and this is the point in which the auxiliary plane cuts the directrix PQ. OX will then be the required element. If the curve $r'w'$ should intersect $p'q'$ in more than one point, two or more elements passing through O would thus be determined.

84. *If an element be required parallel to a given right line*, this line being in the plane directer, or parallel to it, we draw through the different points of either directrix, lines parallel to the given line. These form the rectilinear elements of a cylinder, parallel to the given line. If the points in which the second directrix pierces this cylinder be found, and lines be drawn through them parallel to the given line, each will touch both directrices, and be an element required.

Construction. Let the surface be given as in the preceding Article, and let FS, Fig. 44, be the given line. Through the points O, K, L, &c., draw OX, KY, LZ, &c., parallel to FS. These lines pierce the horizontal projecting cylinder of the directrix PQ in the points X, Y, Z, &c., which, being joined, form the curve XY, intersecting PQ in W. Through W draw WR, parallel to FS; it is a required element, and there may be two or more as in the preceding Article.

85. A warped surface, with a plane directer, having one directrix a right line and the other a curve, is called a *Conoid*. The elements of the conoid pass through all the points of the rectilinear directrix, instead of a single point, as in the cone.

If the rectilinear directrix is perpendicular to the plane directer, it is *a right conoid*, and this directrix is *the line of striction.*

THE HYPERBOLIC PARABOLOID.

86. A warped surface with *a plane directer* and *two rectilinear directrices*, is *a Hyperbolic Paraboloid*, since its intersection by a plane may be proved to be either an hyperbola or parabola.

Its rectilinear elements may be constructed by the principles in Arts. (83 and 84). In the first case, two right lines through the given point will determine the auxiliary plane, and the point in which it is pierced by the second directrix may be determined at once, as in Art. (41). In the second case, the cylinder becomes a plane, and the point in which it is pierced by the second directrix is also determined as in Art. (41).

87. *The rectilinear elements of a hyperbolic paraboloid divide the directrices proportionally;* for these elements are in a system of planes parallel to the plane directer and to each other, and these planes divide the directrices into proportional parts at the points where they are intersected by the elements. (*Davies' Legendre*, Book vi., Prop. xv.)

Conversely. *If two right lines be divided into any number of proportional parts*, the right lines joining the corresponding points of division will lie in a system of parallel planes, and *be elements of an hyperbolic paraboloid*, the plane directer of which is parallel to any two of these dividing lines.

Thus, let MN and OP, Fig. 45, be any two rectilinear directrices. Take any distance, as *ml*, and lay it off on *mn* any number of times, as *ml*, *lk*, *kg*, &c. Take also any distance, as *pr*, and lay it off on *po* any number of times, as *pr*, *rs*, *su*, *qo*, &c. Join the corresponding points of division by right lines, *lr*, *ks*, *gu*, &c.; these will be the horizontal projections of rectilinear elements of the surface. Through *m*, *l*, *k*, &c., and *p*, *r*, *s*, &c., erect perpendiculars to AB, to *m′*, *l′*, *k′*, &c., and *p′*, *r′*, *s′*, &c., and join the corresponding points by the right lines $l'r'$, $k's'$, $g'u'$ &c., these will be the vertical projections of the elements.

88. *To assume any point on this surface*, and in general, on any warped surface, we first assume its horizontal projection, and at this, erect a perpendicular to the horizontal plane. Through this perpendicular pass a plane; it will intersect the rectilinear elements in points which, joined, will give a line of the surface, and the points in which this line intersects the perpendicular will be the required points on the surface.

Let the construction be made upon either of the figures, 43, 44, or 45.

89. *If any two rectilinear elements of an hyperbolic paraboloid be taken as directrices, with a plane directer parallel to the first directrices*, and a surface be thus generated, it will be identical with the first surface.

To prove this, we have only to prove that any point of an element of the *second generation* is also a point of *the first*.

Thus, let MN and OP, Fig. 46, be the directrices of the first generation, and NO and MP any two rectilinear elements. Through M draw MW parallel to NO. The plane WMP will be parallel to the plane directer of the first generation, and may be taken for it.

Let NO and MP be taken as the new directrices, and let ST be an element of the second generation, the plane directer being parallel to MN and OP. Through U, any point of ST, pass a plane parallel to WMP, cutting the directrices MN and OP in Q and R. Join QR, it will be an element of the first generation Art. (83). Draw Nn and Qq parallel to ST, piercing the plane WMP in n and q. Also draw Oo and Rr parallel to ST, piercing WMP in o and r; and draw no, intersecting MP in T, since Nn ST, and Oo are in the same plane. M, q, and n will be in the same right line, as also P, r, and o; and since MN and Nn, intersecting at N, are parallel to the plane directer of the second generation, their plane will be parallel to it, as also the plane of PO and Oo; hence, these planes being parallel, their intersections, Mn and Po, with the plane WMP, will be parallel. Draw qr.

Since Qq and Nn are parallel, we have

$$MQ : QN :: Mq : qn.$$

Also,

$$PR : RO :: Pr : ro.$$

But Art. (87),

$$MQ : QN :: PR : RO;$$

hence,

$$Mq : qn :: Pr : ro;$$

and the right line, qr, must pass through T, and the plane of the three parallels, Qq, ST, and Rr, contains the element QR, which must therefore intersect ST at U. Hence, any point of a rectilinear element of the second generation is also a point of an element of the first generation, and the two surfaces are identical. It follows from this, that *through any point of an hyperbolic paraboloid two rectilinear elements can always be drawn.*

WARPED SURFACES WITH THREE LINEAR DIRECTRICES.

90. A second class of warped surfaces consists of those which may be generated by moving a right line so as to touch three lines as directrices; or *warped surfaces with three linear directrices.*

A rectilinear element of this class of surfaces passing through a given point on one of the directrices, may be found by drawing through this point a system of right lines intersecting either of the other directrices; these form the surface of a cone which will be pierced by the third directrix in one or more points, through which and the given point right lines being drawn will touch the three directrices, and be required elements.

Construction. Let MN, OP, and QR, Fig. 47, be any three linear directrices, and M a given point on the first. Through M draw the lines MO, MS, MP, &c., intersecting OP in O, S, P, &c. They pierce the horizontal projecting cylinder of QR in X, Y, Z, &c., forming a curve, XYZ, which intersects QR in U, the point in which QR pierces the surface of the auxiliary cone. MU is then a required element, two or more of which would be determined if XZ should intersect QR in more than one point.

91. The simplest of the above class of surfaces is *a surface which may be generated by moving a right line so as to touch three rectilinear directrices.* It is *an Hyperboloid of one nappe*, as many of its intersections by planes are hyperbolas, while the surface itself is unbroken.

The construction of the preceding Article is much simplified for this surface, as the auxiliary cone becomes a plane; and the point in which this plane is pierced by the third directrix is found as in Art. (40), or (41).

THE HELICOID.

92. If a right line be moved uniformly along another right line, as a directrix, always making the same angle with it, and at the same time having a uniform angular motion around it, a warped surface will be generated, called *a Helicoid.*

The rectilinear directrix is *the axis of the surface.* Thus, Fig, 48, let the horizontal plane be taken perpendicular to the axis, *o* being its horizontal, and *o'n'* its vertical projection, and let OP be the generatrix, parallel to the vertical plane.

It is evident from the nature of the motion of the generatrix, that each of its points will generate a helix, Art. (68). That generated by the point P, constructed as in Art. (68), will be horizontally projected in prq, and vertically in $p'r'q'$.

To assume a rectilinear element of the surface, we first assume its horizontal projection, as ox. Through x erect the perpendicular xx', and from o' lay off the distance $o'o''$, equal to $x'x''$; $o''x'$ will be the required vertical projection. In the same way, the element (oy, $o'''y'$) may be assumed.

To assume any point on the surface, we first assume a rectilinear element, and then take any point in this element, as M.

The elements pierce the horizontal plane in the points p, u, w, &c., and the curve puw is its intersection by the surface.

The surface is manifestly a warped surface, since, from the nature of its generation, no two consecutive positions of the generatrix can be parallel or intersect.

This surface is an important one, as it forms the curved surface of the thread of the ordinary screw.

If the generatrix is perpendicular to the axis, the helicoid becomes a particular case of the right conoid, Art. (85), the horizontal plane being the plane directer, and any helix the curvilinear directrix.

93. Other varieties of warped surfaces may be generated by moving a right line so as to touch two lines, having its different positions, in succession, parallel to the different rectilinear elements of a cone: By moving it parallel to a given plane, so as to touch two surfaces, or one surface and a line; or so as to touch three surfaces—two surfaces and a line—one surface and two lines; and in general, so as to fulfil any three reasonable conditions.

It should be remarked, that in all these cases the directrices should be so chosen that the surface generated will be neither a plane nor single curved surface. Also, that these surfaces, as all others, may be generated by curves moved in accordance with a law peculiar to each variety.

94. If a curve be moved in any way so as not to generate a surface of either of the above classes, it will generate *a double curved surface*, the simplest variety of which is *the surface of a sphere.*

95. An examination of the modes of generating surfaces, above described, will indicate the following test for ascertaining the general class to which any given surface belongs.

1. If a straight ruler can be made to touch the surface in every direction, it must be a plane.
2. If the ruler can be made to touch a curved surface in any one direction, and then be moved so as to touch in a position very near to the first, and the two right lines thus indicated are parallel or intersect, the surface must be *single curved.*
3. If the lines thus indicated, neither intersect nor are parallel, the surface must be *warped.*
4. If the ruler cannot be made to touch the surface in any direction, the surface must be *double curved.*

SURFACES OF REVOLUTION.

96. *A surface of revolution, is a surface which may be generated by revolving a line about a right line as an axis,* Art. (17).

From the nature of this generation, it is evident that any intersection of such a surface by a plane perpendicular to the axis, is the circumference of a circle. Hence, the surface may also be generated by moving the circumference of a circle with its plane perpendicular to the axis, and centre in the axis, and whose radius changes in accordance with a prescribed law.

If the surface be intersected by a plane passing through the axis, the line of intersection is *a meridian line*, and the plane *a meridian plane;* and it is also evident that all meridian lines of the same surface are equal, and that the surface may be generated by revolving any one of these meridian lines about the axis

97. If two surfaces of revolution, having a common axis intersect, *the line of intersection must be the circumference of a circle*, whose plane is perpendicular to the axis, and centre in the axis, For if a plane be passed through any point of the intersection and the common axis, it will cut from each surface a meridian line, Art. (96), and these meridian lines will have the point in common. If these lines be revolved about the common axis, each will generate the surface to which it belongs, while the common point will generate the circumference of a circle common to the two surfaces, and therefore their intersection. Should the meridian lines intersect in more than one point, the surfaces will intersect in two or more circumferences.

98. The simplest curved surface of revolution is that which may be generated by a right line revolving about another right line to which it is parallel. This is evidently a cylindrical surface, Art. (73), and if the plane of the base be perpendicular to the axis, it is *a right cylinder with a circular base.*

If a right line be revolved about another right line which it intersects, it will generate a conical surface, Art. (77), which is evidently *a right cone*, the axis being the line with which the rectilinear elements make equal angles.

These are the only two single curved surfaces of revolution.

THE HYPERBOLOID OF REVOLUTION OF ONE NAPPE.

99. If a right line be revolved about another right line, not in the same plane with it, it will generate a warped surface, which is *an Hyperboloid of revolution of one nappe*, Art. (91).

To prove this, let us take the horizontal plane perpendicular to the axis, and the vertical plane parallel to the generatrix in its first position, and let c, Fig. 49, be the horizontal, and $c'm'$ the vertical projection of the axis, and MP the generatrix: cm will be the horizontal, and m' the vertical projection of the shortest distance between these two lines, Art. (55).

As MP revolves about the axis, CM will remain perpendicular to it, and M will describe a circumference which is horizontally projected in mxy, and vertically in $y'x'$; and as CM is horizontal, its horizontal projection will be perpendicular to the horizontal projection of MP, in all of its positions, Art. (36), and remain of the same length; hence the horizontal projection of MP, in any position, will be tangent to the circle mxy. No two consecutive positions can therefore be parallel. Neither can they intersect; for from the nature of the motion, any two must be separated at any point by the elementary arc of the circle described by that point. *The surface is therefore a warped surface.*

The point P, in which MP pierces H, generates the circle pwq, which may be regarded as *the base* of the surface.

The circle generated by M, is the smallest circle of the surface, and is *the circle of the gorge.*

100. *To assume a rectilinear element,* we take any point in the base, pwq, and through it draw a tangent to xmy, as wz; this will be the horizontal projection of an element. Through z erect the perpendicular zz'; z' will be the vertical projection of the point

in which the element crosses the circle of the gorge, and $w'z'$ will be the vertical projection of the element.

To assume a point of the surface, we first assume a rectilinear element, as above, and then take any point of this element, Art. (22).

101. If through the point M, a second right line, as MQ, be drawn parallel to the vertical plane, and making with the horizontal plane the same angle as MP, and this line be revolved about the same axis, it will generate the same surface. For if any plane be passed perpendicular to the axis, as the plane whose vertical trace is $e'g'$, it will cut MP and MQ in two points, E and G, equally distant from the axis, and these points will, in the revolution of M P and MQ, generate the same circumference; hence the two surfaces must be identical. The surface having two generations by different right lines, it follows that *through any point of the surface two rectilinear elements can be drawn.*

102. Since the points E and G generate the same circumference, it follows that as MP revolves, MQ remaining fixed, the point E will, at some time of its motion, coincide with G, and the generatrix, MP, intersect MQ in G. In the same way any other point, as F, will come into the point K of MQ, giving another element of the first generation, intersecting MQ at K; and so for each of the points of MP in succession. In this case, kl will be the horizontal projection of the element of the first generation. Hence, *if the generatrix of either generation remain fixed, it will intersect all the elements of the other generation.*

If, then, any three elements of either generation be taken as directrices, and an element of the other generation be moved so as to touch them, it will generate the surface. It is therefore *an hyperboloid of one nappe*, Art. (91).

An hyperboloid of revolution of one nappe may thus be gen

erated, *by moving a right line so as to touch three other right lines*, equally distant from a fourth (the axis), the distances being measured in the same plane perpendicular to the fourth, and the directrices making equal angles with this plane.

103. If the vertical plane is not parallel to the generatrix in its first position, the circle of the gorge may be constructed thus: Let WZ be the first position of the generatrix, and through *c* draw *cz* perpendicular to *wz*; it will be the horizontal projection of the radius of the required circle. The point of which *z* is the horizontal projection, is vertically projected at *z'*, and *mzx* will be the horizontal, and *y'x'* the vertical projection of the circle of the gorge. With *c* as a centre, and *cw* as a radius, describe the base *pwq*, and the surface will be fully represented.

104. *To construct a meridian curve* of this surface, we pass a plane through the axis, parallel to the vertical plane. It will intersect the horizontal circles, generated by the different points of the generatrix, in points of the required curve, which will be vertically projected into a curve equal to itself, Art. (62).

Thus the horizontal plane, whose vertical trace is *e'g'*, intersects the generatrix in E, and *eo* is the horizontal projection of the circle generated by this point, and O is the point in which this circle pierces the meridian plane.

In the same way, the points whose vertical projections are *n'*, *x'*, *n''*, *o''*, &c., are determined. The plane whose vertical trace is *e''o''*, at the same distance from *y'x'* as *e'o'*, evidently determines a point *o''* at the same distance from *y'x'*, as *o'*; hence the chord *o'o''* is bisected by *y'x'*, and the curve *o'x'o''* is symmetrical with the line *y'x'*.

The distance *e'o'* equal to *ce* — *me*, is the difference between the hypothenuse *ce* and base *me* of a right-angled triangle, having the altitude *mc*. As the point *o'* is further removed from *x'*, the altitude of the corresponding triangle remains the same, while the hypothenuse and base both increase. If we denote the altitude by a, and the base and hypothenuse by b and h respectively, we have

$$h^2 - b^2 = a^2, \quad \text{and} \quad h - b = \frac{a^2}{h + b};$$

from which it is evident that the difference, $e'o'$, continually diminishes as the point o' recedes from x'; that is, the curve $x'n'o$ continually approaches the lines $p'm'$ and $q'm'$, and will touch them at an infinite distance. These lines are then asymptotes to the curve, Art. (64).

If, now, any element of the first generation, as the one passing through I, be drawn, it will intersect the element of the second generation, MQ, in R, and the corresponding element on the opposite side of the circle of the gorge in U.

Since this element has but one point in common with the meridian curve, and no point of it, or of the surface, can be vertically projected on the right of this curve, the vertical projection $u'r'$ must be tangent to the curve at i'. But since ri is equal to iu, $r'i'$ must be equal to $i'u'$; or the part of the tangent included between the asymptotes *is bisected at the point of contact.* This is a property peculiar to the hyperbola (*Analyt. Geo.*, Art. 164), and the meridian curve is therefore an hyperbola; YX being its transverse axis, and the axis of the surface its conjugate, Art (59). If this hyperbola be revolved about its conjugate axis, it will generate the surface, Art. (96), and hence its name.

This is the only warped surface of revolution.

105. The most simple double curved surfaces of revolution are:

I. *A spherical surface or sphere,* which may be generated by revolving a circumference about its diameter.

II. *An ellipsoid of revolution, or spheroid,* which may be generated by revolving an ellipse about either axis. When the ellipse is revolved about the transverse axis, the surface is *a prolate spheroid ;* when about the conjugate axis, *an oblate spheroid.*

III. *A paraboloid of revolution,* which may be generated by revolving a parabola about its axis.

IV. *An hyperboloid of revolution of two nappes,* which may be generated by revolving an hyperbola about its transverse axis.

106. These surfaces of revolution are usually represented by taking the horizontal plane perpendicular to the axis, and then drawing the intersection of the surface with the horizontal plane; or the horizontal projection of the greatest horizontal circle of the surface for the horizontal projection, and then projecting on the vertical plane the meridian line which is parallel to that plane, for the vertical projection.

107. *To assume a point on a surface thus represented,* we first assume either projection, as the horizontal, and erect a perpendicular to the horizontal plane, as in Art. (76). Through this perpendicular pass a meridian plane; it will cut from the surface a meridian line, which will intersect the perpendicular in the required point or points.

Construction. Let the surface be a prolate spheroid, Fig. 50, and let c be the horizontal, and $c'd'$ the vertical projection of the axis, mon the horizontal projection of its largest circle, and $d'm'c'n'$ the vertical projection of the meridian curve parallel to the vertical plane, and let p be the assumed horizontal projection ; p will be the horizontal, and $s'p'$ the vertical projection of the perpendicular to H : cp will be the horizontal trace of the auxiliary meridian plane. If this plane be now revolved about the axis until it becomes parallel to the vertical plane, p will describe the arc pr, and r will be the horizontal, and $r'r''$ the vertical projection of the revolved position of the perpendicular. The meridian curve, in its revolved position, will be vertically projected into its equal, $d'm\,c'n'$, Art. (61), and r' and r'' will be the vertical projections of the two points of intersection in their revolved position. When the meridian plane is revolved to its primitive position, these points will describe the arcs of horizontal circles, projected on H in rp, and on V in $r'p'$ and $r''p''$, and p' and p'' will be the vertical projections of the required points of the surface.

TANGENT PLANES AND TANGENT SURFACES. NORMAL LINES AND NORMAL PLANES.

108. *A plane is tangent to a surface when it has at least one point in common with the surface, through which, if any intersecting plane be passed, the right line cut from the plane will be tangent to the line cut from the surface at the point.* This point is *the point of contact.*

It follows from this definition, that the tangent plane is *the locus of, or place in which are to be found, all right lines tangent to lines of the surface at the point of contact;* and since any two of these right lines are sufficient to determine the tangent plane, we have the following general rule for passing a plane tangent to any surface at a given point: *Draw any two lines of the surface intersecting at the point. Tangent to each of these, at the same point, draw a right line.* The plane of these two tangents *will be the required plane.*

109. A right line *is normal to a surface*, at a point, when *it is perpendicular to the tangent plane at that point.*

A plane is *normal to a surface*, when *it is perpendicular to the tangent plane at the point of contact.* Since any plane passed through a normal line will be perpendicular to the tangent plane there may be an infinite number of normal planes to a surface at a point, while there can be but one normal line.

110. The tangent plane at any point of a surface with recti linear elements, *must contain the rectilinear elements that pass through the point of contact.* For the tangent to each rectilinear element is the element itself, Art. (64), and this tangent must lie in the tangent plane, Art. (108).

111. A plane which *contains a rectilinear element and its consecutive one*, of a single curved surface, will be tangent to the surface at every point of this element, or *all along the element.* For if through *any point* of the element *any intersecting plane* be passed, it will intersect the consecutive element in a point consecutive with the first point. The right line joining these two points will lie in the given plane, and be tangent to the line cut from the surface, Art. (64). Hence the given plane will be tangent at the assumed point.

This element is *the element of contact.*

Conversely, if a plane be tangent to a single curved surface, it must, in general, *contain two consecutive rectilinear elements.* For if through any point of the element contained in the tangent plane, Art. (110), we pass an intersecting plane, it will cut from the surface a line, and from the plane a right line, which will have two consecutive points in common, Art. (108). Through the point consecutive with the assumed point, draw the consecutive element to the first element; it must lie in the plane of the second point and first element, Art. (70), that is, in the tangent plane.

112. It follows from these principles, that if a plane be tangen to a single curved surface, and the element of contact be interected by any other plane, the right line cut from the tangent plane will be tangent to the line cut from the surface. Hence, if the base of the surface be in the horizontal plane, *the horizontal trace of the tangent plane must be tangent to this base*, at the point

in which the element of contact pierces the horizontal plane; and the same principle is true if the base lie in the vertical plane.

113. A plane tangent to a warped surface, although it contains the rectilinear element passing through the point of contact, cannot contain its consecutive element, and therefore can, in general, be tangent at no other point of the element.

114. If a plane *contain a rectilinear element of a warped surface*, and be not parallel to the other elements, *it will be tangent to the surface at some point of this element.* For this plane will intersect each of the other rectilinear elements in a point; and these points being joined, will form a line which will intersect the given element. If at the point of intersection a tangent be drawn to this line, it will lie in the tangent plane, Art. (108). The given element being its own tangent, Art. (64), also lies in the tangent plane. The plane of these two tangents, that is, the intersecting plane, is therefore tangent to the surface at this point, Art. (108). Thus, Fig. 51, if a is an element, b, c, &c., b', c', &c., the consecutive elements on each side of a, the plane through a will cut these elements in the curve $cbab'c'$, which crosses the element a at the point a. The tangent mn at this point, and the element a, determine the tangent plane. It is thus seen, that, in general, a tangent plane to a warped surface is also *an intersecting plane.*

If the intersecting plane be parallel to the rectilinear elements, there will be no curve of intersection formed as above, and the plane will not be tangent.

115. A plane tangent to a surface of revolution, *is perpendicular to the meridian plane* passing through the point of contact. For this tangent plane contains the tangent to the circle of

the surface at this point, Art. (108), and this tangent is perpendicular to the radius of this circle, and also to a line drawn through the point parallel to the axis, since it lies in a plane perpendicular to the axis; and therefore it is perpendicular to the plane of these two lines, which is the meridian plane.

116. Two curved surfaces are tangent to each other when they have at least one point in common, through which if any intersecting plane be passed, the lines cut from the surfaces will be tangent to each other at this point. This will evidently be the case when they have a common tangent plane at this point.

117. If two single curved surfaces are tangent to each other, at a point of a common element, they will be tangent *all along this element.* For the common tangent plane will contain this element, and be tangent to each surface at every point of the element, Art. (111).

This principle is not true with warped surfaces.

118. But if two warped surfaces, having two directrices, *have a common plane directer, a common rectilinear element, and two common tangent planes, the points of contact being on the common element,* they will be tangent all along this element. For if through each of the points of contact any intersecting plane be passed, it will intersect the surfaces in two lines, which will have, besides the given point of contact, a second consecutive point in common, Art. (108). If, now, the common element be moved pon the lines cut from either surface, as directrices, and parallel to the common plane directer, into its consecutive position, containing these second consecutive points, it will evidently lie in both surfaces, and the two surfaces will thus contain two consecutive rectilinear elements. If, now, *any plane* be passed, intersect-

ing these elements, two lines will be cut from the surfaces, having two consecutive points in common, and therefore tangent to each other; hence the surfaces will be tangent all along the common element, Art. (116).

119. Also, if two warped surfaces, having three directrices, *have a common element and three common tangent planes, the points of contact being on this element*, they will be tangent all along this element. For if through each point of contact any intersecting plane be passed, it will intersect the surfaces in two lines, which, besides the given point of contact, will have a second consecutive point in common. If the common element be moved upon the three lines cut from either surface, as directrices, to its consecutive position, so as to contain the second consecutive points, it will evidently lie in both surfaces; hence the two surfaces contain two consecutive rectilinear elements, and will be tangent all along the common element.

120. It follows from the principle in Art. (114), that the projecting plane of every rectilinear element of a warped surface is tangent to the surface at a point. If through these points of tangency, projecting lines be drawn, they will form a cylinder tangent to the warped surface, which may be regarded as the projecting cylinder of the surface; and the traces of these planes, or the projections of the elements, will all be tangent to the base of this cylinder. This is seen in the horizontal projection of the hyperboloid of revolution of one nappe, Fig. 49; also in both projections of the hyperbolic paraboloid, Fig. 45.

121. If two surfaces of revolution, having a common axis, are tangent to each other, they will be *tangent at every point of a circumference of a circle*, perpendicular to the axis. For if

through the point of contact a meridian plane be passed, it will cut from the surfaces meridian lines, which will be tangent at this point, Art. (108). If these lines be revolved about the common axis, each will generate the surface to which it belongs, while the point of tangency will generate a circumference common to the two surfaces, which is their line of contact.

SOLUTION OF PROBLEMS RELATING TO TANGENT PLANES TO SINGLE CURVED SURFACES.

122. The solution of all these problems depends mainly upon the principles, that a plane tangent to a single curved surface is tangent all along a rectilinear element, Art. (111); and that if such surface and tangent plane be intersected by any plane, the lines of intersection will be tangent to each other.

123. PROBLEM 18. *To pass a plane tangent to a cylinder at a given point on the surface.*

Let the cylinder be given as in Art. (75), Fig. 40, and let P be the point, assumed as in Art. (76).

Analysis. Since the required plane must contain the rectilinear element through the given point, and its horizontal trace must be tangent to the base at the point where this element pierces the horizontal plane, Art. (112), we draw the element, and at the point where it intersects the base, a tangent; this will be the horizontal trace. The vertical trace must contain the point where this element pierces the vertical plane, and also the point where the horizontal trace intersects the ground line. A right line joining these two points will be the vertical trace. When this element does not pierce the vertical plane, within the limits of the drawing, we draw through any one of its points a line parallel to the horizontal trace; it will be a line of the required plane, and pierce the vertical plane in a point of the vertical trace.

Construction. Draw the element PQ; it pierces H at q. At this point draw qT, tangent to mlo; it is the required horizontal trace. Through P draw PZ, parallel to qT; it pierces V at z', and z'T is the vertical trace.

$mm'n'$ is the plane tangent to the surface along the element MN.

124. PROBLEM 19. *To pass a plane through a given point, without the surface, tangent to a cylinder.*

Let the cylinder be given as in the preceding problem.

Analysis. Since the plane must contain a rectilinear element; if we draw a line through the given point, parallel to the rectilinear elements of the cylinder, it must lie in the tangent plane, and the point in which it pierces the horizontal plane, will be one point of the horizontal trace. If through this point we draw a tangent to the base, it will be the required horizontal trace. A line through the point of contact, parallel to the rectilinear elements, will be the element of contact; and the vertical trace may be constructed as in the preceding problem: Or a point of this trace may be obtained by finding the point in which the auxiliary line pierces the vertical plane.

Two or more tangent planes may be passed, if two or more tangents can be drawn to the base, from the point in which the auxiliary line pierces the horizontal plane.

Let the construction be made in accordance with the above analysis.

125. PROBLEM 20. *To pass a plane, tangent to a cylinder, and parallel to a given right line.*

Let the cylinder be given as in Fig. 52, and let RS be the given line.

Analysis. Since the required plane must be parallel to the rectilinear elements of the cylinder, as well as to the given line; if a plane be passed through this line, parallel to a rectilinear element, it will be parallel to the required plane, and its traces parallel to the required traces, Art. (10). Hence, a tangent to the base, parallel to the horizontal trace of this auxiliary plane, will be the required horizontal trace. The element of contact and vertical trace may be found as in the preceding problem.

Construction. Through RS pass the plane $sT'r'$, parallel to MN, as in Art. (48). Tangent to mlo and parallel to sT', draw qT; it is the horizontal trace of the required plane, and Tt' parallel to $T'r'$ is the vertical trace. QP is the element of contact. The point z', determined as in the preceding problem, will aid in verifying the accuracy of the drawing.

When more than one tangent can be drawn parallel to sT', there will be more than one solution to the problem.

126. PROBLEM 21. *To pass a plane tangent to a right cylinder with a circular base, having its axis parallel to the ground line at a given point on the surface.*

Let cd, Fig. 53, be the horizontal, and $e'f'$ the vertical projection of the circular base, and GK the axis; then $lcdi$ will be the horizontal, and $u'e'f'b'$ the vertical projection of the cylinder. Let p be the horizontal projection of the point. To determine its vertical projection, at p erect a perpendicular to H, Art. (76). Through this pass the plane tTt' perpendicular to AB. It intersects the cylinder in the circumference of a circle equal to the base, of which M is the centre. This circumference will intersect the perpendicular in two points of the surface. Revolve this plane about tT until it coincides with H; M falls at m'' Art. (17). With m'' as a centre and cg as a radius, describe the circle qsr; it will be the revolved position of the circle cut from the cylinder; q and r will be the revolved positions of the required points, and p' and p'' their vertical projections. Let the plane be passed tangent at P.

Analysis. Since the plane must contain a rectilinear element, it will be parallel to the ground line, and its traces, therefore parallel to the ground line, Art. (9). If through the given point a plane be passed perpendicular to the axis, and a tangent be drawn to its intersection with the cylinder, at the point, it will be a line of the required plane. If through the points in which this line pierces the planes of projection, lines be drawn parallel to the ground line, they will be the required traces.

Construction. Through P pass the plane tTt', and revolve it as above. At q draw qx tangent to qrs; it is the revolved position of the tangent. When the plane is revolved to its primitive position, this tangent pierces H at x, and V at y', Ty' being equal to Ts'; and xz and $y'w'$ are the required traces.

127. Problem 22. *To pass a plane tangent to a right cylinder with a circular base, having its axis parallel to the ground line, through a given point without the surface.*

Let the cylinder be given as in Fig. 53, and let N be the given point.

Analysis. Since the required plane must contain a rectilinear element, it must be parallel to the ground line. If through the given point a plane be passed perpendicular to the axis, it will cut from the cylinder a circumference equal to the base; and if through the point a tangent be drawn to this circumference, it will be a line of the required plane, and the traces may be determined as in the preceding problem.

Since two tangents can be drawn, there may be two tangent planes.

Let the construction be made in accordance with the analysis.

128. Problem 23. *To pass a plane parallel to a given right line, and tangent to a right cylinder, with a circular base, with its axis parallel to the ground line.*

Let the cylinder be given as in Fig. 54, and let MN be the given line.

Analysis. Since the plane must be parallel to the axis, if through the given line we pass a plane parallel to the axis, it will be parallel to the required plane, and to the ground line.

A plane perpendicular to the ground line will cut from the cylinder a circumference; from the required tangent plane a tangent to this circumference, Art. (108); and from the parallel plane a right line parallel to the tangent. If, then, we construct the circumference, and draw to it a tangent parallel to the intersection of the parallel plane, this tangent will be a line of the required plane, from which the traces may be found as in the preceding problems.

Construction. Through m and n' draw the lines mp and $n'q'$, parallel to AB. They will be the traces of the parallel plane, Art. (48). Let tTt' be the plane perpendicular to AB. It cuts from the cylinder a circle whose centre is O, and from the parallel plane a right line which pierces H at p and V at q', Art. (38). Revolve this plane about tT, until it coincides with H; xyz will be the revolved position of the circle, and sp that of the line cut from the parallel plane; zu tangent to xyz, and parallel to sp, will be the revolved position of a line of the required plane, which, in its true position, pierces H at u, and V at w', and uv and $w'v'$ are the traces of the required plane.

Since another parallel tangent can be drawn, there will be two solutions.

129. Problem 24. *To pass a plane tangent to a cone, through a given point on the surface.*

Let the cone be given as in Fig. 41, and let P, assumed as in Art. (79), be the given point.

Analysis. The required plane must contain the rectilinear element, passing through the given point, Art. (110). If, then, we draw this element, Art. (79), and at the point where it pierces the horizontal plane, draw a tangent to the base, it will be the horizontal trace of the required plane, and points of the vertical trace may be found as in the similar case for the cylinder.

Let the construction be made in accordance with the analysis.

130. PROBLEM 25. *Through a point without the surface of a cone, to pass a plane tangent to the cone.*

Let the cone be given as in Fig. 55, and let P be the given point.

Analysis. Since the required plane must contain a rectilinear element, it will pass through the vertex; hence, if we join the given point with the vertex by a right line, it will be a line of the required plane, and pierce the horizontal plane in a point of the required horizontal trace, which may then be drawn tangent to the base. If the point of contact with the base be joined to the vertex by a right line, it will be the element of contact. The vertical trace may be found as in the preceding problems.

If more than one tangent can be drawn to the base, there will be more than one solution.

Construction. Join P with S, by the line PS; it pierces H at u. Draw uq tangent to mlo; it is the required horizontal trace. SQ is the element of contact which pierces V at w', and z'T is the vertical trace.

ux will be the horizontal trace of a second tangent plane through P.

131. PROBLEM 26. *To pass a plane tangent to a cone, and parallel to a given right line.*

Let the cone be given as in Fig. 55, and let NR be the given right line.

Analysis. If through the vertex we draw a line parallel to the given line, it must lie in the required plane, and pierce the horizontal plane in a point of the horizontal trace. Through this point draw a tangent to the base, it will be the required horizontal trace; and the element of contact and vertical trace may be found as in the preceding problem.

When more than one tangent can be drawn to the base, there will be more than one solution.

If the parallel line through the vertex pierces the horizontal plane within the base, no tangent can be drawn, and the problem is impossible.

Let the construction be made in accordance with the analysis.

132. Problem 27. *To pass a plane tangent to a single curved surface with a helical directrix, at a given point.*

Let the surface be given as in Fig. 42, and let R be the given point, Art. (81).

Analysis. The tangent plane must, in general, be tangent all along the rectilinear element, through the point of contact. Its horizontal trace must therefore be tangent to the base, Art. (112). If through the point where this element pierces the horizontal plane, a tangent be drawn to the base, it will be the required horizontal trace, and the vertical trace may be determined as in the case of the cylinder or cone.

Construction. The element RX pierces H at Z. At this point draw the tangent tT; it is the horizontal trace. Tt' is the vertical trace.

To pass a plane through a given point without this surface, tangent to it, we pass a plane through the point parallel to the base, and draw a tangent to the curve of intersection, Art. (81), through the point. This tangent, with the element of the surface through its point of contact, will determine the tangent plane.

133. Problem 28. *To pass a plane tangent to a single curved surface with a helical directrix, and parallel to a given right line.*

Let the surface be given as in Fig. 42, and let MN be the given line.

Analysis. If with any point of the right line, as a vertex, we construct a cone, whose elements make the same angle with the horizontal plane as the elements of the surface; and pass a plane through the line tangent to this cone, it will be parallel to the required plane. The traces of the required plane may then be constructed as in Art. (125).

Construction. Take n' as the vertex of the auxiliary cone, and draw $n'o''$, making with AB an angle equal to $o's'$T; $o''cs''$ will be the base of the cone in H. Through m draw mc tangent to $o''cs''$. It will be the horizontal trace of the parallel plane, and t''T', parallel to it, and tangent to uvw, is the required horizontal trace. Let the pupil construct the vertical trace.

134. It is a remarkable property of this surface, with a helical directrix, *that any plane passing through a rectilinear element, is tangent to the surface*, at the point where the element intersects the directrix. For this plane will intersect the other elements, thus forming a curve, Art. (114). This curve will intersect the given element at the point where the element touches the directrix. The tangent to the curve at this point, and the element, both lie in the given plane; it is therefore tangent to the surface at the point, Art. (114). This plane does not, in general, contain the consecutive element, and is therefore not necessarily tangent all along the element.

The projecting planes of all the elements are tangent to the surface, and the cylinder formed by the projecting lines of the points of contact is therefore tangent to the surface, Art. (120). Its base is the circle pxq.

It should be observed, also, that at any point on the helical directrix, an infinite number of planes can be passed tangent to the surface, as at the vertex of a cone. Only one of these planes will contain two consecutive rectilinear elements.

135. By an examination of the preceding problems, it will be seen that, with two remarkable exceptions, only one tangent plane can be drawn to a single curved surface at a given point.

That the number which can be drawn through a given point without the surface, and tangent along an element, will be limited.

That the number which can be drawn parallel to a given right line, and tangent along an element, is also limited.

That, in general, a plane cannot be passed through a given right line and tangent to a single curved surface. If, however, the given line lies on the convex side of the surface, and is parallel to the rectilinear elements of a cylinder, or passes through the vertex of a cone, or is tangent to a line of the surface, the problem is possible.

PROBLEMS RELATING TO TANGENT PLANES TO WARPED SURFACES.

136. Since a tangent plane to a warped surface must contain the rectilinear element passing through the point of contact, Art. (110), we can at once determine one line of the tangent plane. A second line may then be determined, in accordance with the rule in Art. (108).

When the surface has two different generations by right lines, the plane of the two rectilinear elements passing through the given point will be the required plane.

137. Problem 29. *To pass a plane tangent to a hyperbolic paraboloid, at a given point of the surface.*

Let MN and PQ, Fig. 56, be the directrices, and MP and NQ any two elements of the surface, and let O, assumed as in Art. (88), be the given point.

Analysis. Since through the given point a rectilinear element of each generation can be drawn, Art. (89), we have simply to construct these two elements, and pass a plane through them, Art. (136).

Construction. Through O draw OE and OF, parallel respectively to NQ and MP. These will determine a plane parallel to the plane directer of the first generation, Art. (86). This plane cuts the directrix PQ in the point U, Art. (41). Join U with O and we have an element of the first generation, Art. (83). Let NQ and MP be taken as directrices of the second generation

Through O draw OC and OD parallel respectively to MN and PQ. They will determine a plane parallel to the plane directer of the second generation, Art. (86). This plane cuts MP in W, and OW will be an element of the second generation, and the tangent plane is determined, as in Art. (32).

138. Problem 30. *To pass a plane tangent to an hyperboloid of revolution of one nappe, at a given point of the surface.*

Let the surface be given as in Fig. 57, and let O be the given point.

Analysis. Same as in the preceding problem.

Construction. OZ is the element of the first generation passing through O, Art. (100). It pierces H at *r*. Draw OS, Art. (101). It is an element of the second generation passing through O. This element pierces H at *w*, and *wr* is the horizontal trace of the required plane, and T*t'* is its vertical trace.

Since the meridian plane through O must be perpendicular to the tangent plane, Art. (115), its trace *cx* must be perpendicular to *wr*.

139. Problem 31. *To pass a plane tangent to a helicoid, at a point of the surface.*

Let the surface be given as in Fig. 48, and let M be the given point.

Analysis. The tangent plane must contain the rectilinear element passing through the given point, and also the tangent to the helix at this point, Art. (108). The plane of these two lines will then be the required plane.

Construction. MX is the rectilinear element through M. It pierces H at *u*; *cmd* is the horizontal projection of the helix through M. Draw the tangent to this helix at M, as in Art. (69). *mz* will be its horizontal projection; Z the point in which it pierces the horizontal plane through C, and *m'z'* its vertical projection. This tangent pierces H at k; hence ku is the horizontal trace of the required plane, and t'T is the vertical trace.

Since a tangent plane to the surface contains a rectilinear element, it is evident that it cannot make a less angle with the horizontal plane than the elements; nor a greater angle than 90° the angle made by the projecting plane of any rectilinear element, which is tangent at the point where the element intersects the axis.

140. PROBLEM 32. *To pass a plane tangent to a helicoid, and perpendicular to a given right line.*

Analysis. If, with any point of the axis as a vertex, we construct a cone whose rectilinear elements shall make with the horizontal plane the same angle as that made by the rectilinear elements of the given surface, and through the vertex of this cone pass a plane perpendicular to the given line, Art. (46), it will, if the problem be possible, cut from the cone two elements, each of which will be parallel to a rectilinear element of the helicoid, and have the same horizontal projection. If through either of these elements a plane be passed parallel to the auxiliary plane, it will be tangent to the surface, Art. (114), and perpendicular to the given line.

Let the problem be constructed in accordance with the analysis.

141. An auxiliary surface may sometimes be used to advantage in passing a plane tangent to a warped surface at a given point. Thus, let MN and PQ, Fig. 58, be the two directrices of a warped surface having V for its plane directer, and let O be the given point. At the points X and Y, in which the rectilinear element through O intersects the directrices, draw a tangent to each directrix, as XZ and YU. On these tangents as directrices, move XY parallel to V. It will generate an hyperbolic paraboloid, Art. (86), having, with the given surface, the common element XY,

and a common tangent plane at each of the points X and Y, since the plane of the two lines XY and XZ, and also that of XY and YU, is tangent to both surfaces, Art. (108). The two surfaces are therefore tangent all along the common element, XY, Art. (118) If, then, at O, we pass a plane tangent to the hyperbolic paraboloid, it will be also tangent to the given surface at the same point.

If the given surface have three curvilinear directrices, a tangent may be drawn to each, at the point in which the rectilinear element through the given point intersects it; and then this element may be moved on these three tangents as directrices, generating a hyperboloid of one nappe, Art. (91), which will be tangent to the given surface all along a common element, Art. (119). A plane tangent to this auxiliary surface at the given point, will also be tangent to the given surface.

142. An infinite number of planes may, in general, be passed through a point without a warped surface, and tangent to it. For if, through the point, a system of planes be passed intersecting the surface, tangents may be drawn from the point to the curves of intersection, and these will form the surface of a cone tangent to the warped surface. Any plane tangent to this cone will be tangent to the warped surface, and pass through the point.

Also, an infinite number of planes may. in general, be passed tangent to a warped surface, and parallel to a right line. For if the surface be intersected by a system of planes parallel to the line, and tangents parallel to the line be drawn to the sections, they will form the surface of a cylinder, tangent to the warped surface. Any plane tangent to this cylinder will be tangent to the warped surface, and parallel to the given line.

143. *To pass a plane through a given right line, and tangent to a warped surface*, it is only necessary to produce the line until

it pierces the surface, and through the point thus determined draw the rectilinear element of the surface. This, with the given line, will determine a plane tangent to the surface at some point of the element, Art. (114).

If there be two rectilinear elements passing through this point, each will give a tangent plane; and the number of tangent planes will depend upon the number of points in which the line pierces the surface. If the given line be parallel to a rectilinear element, it pierces the surface at an infinite distance, and the tangent plane will be determined by the two parallel lines.

PROBLEMS RELATING TO TANGENT PLANES TO DOUBLE CURVED SURFACES.

144. These problems are, in general, solved either by a direct application of the rule in Art. (108), taking care to intersect the surface by planes, so as to obtain the two simplest curves of the surface intersecting at the point of contact, or by means of more simple auxiliary surfaces tangent to the given surface.

145. PROBLEM 33. *To pass a plane tangent to a sphere, at a given point.*

Let C, Fig. 59, be the centre of the sphere; prq its horizontal, and $p's'q'$ its vertical projection. Let M be the point assumed, as in Art. (107).

Analysis. Since the radius of the sphere, drawn to the point of contact, is perpendicular to the tangent to any great circle a this point, and since these tangents all lie in the tangent plane, Art. (108), *this radius must be perpendicular to the tangent plane.* We have then simply to pass a plane perpendicular to this radius at the given point, and it will be the required plane.

Construction. This may be made directly, as in Art. (46), or otherwise, thus: Through M and C pass a plane perpendicular to H; *cm* will be its horizontal trace. Revolve this plane about *cm* as an axis, until it coincides with H; *c''m''* will be the revolved position of the radius, Art. (28). At *m''* draw *m''t* perpendicular to *c''m''*. It is the revolved position of a line of the required plane. It pierces H at *t*, and since the horizontal trace must be perpendicular to *cm*, Art. (43). *t*T is the required horizontal trace. Through M draw MN parallel to *t*T. It pierces V at *n'*; and T*n'*, perpendicular to *c'm'*, is the vertical trace.

146. PROBLEM 34. *To pass a plane tangent to an ellipsoid of revolution, at a given point.*

Let the surface be given as in Art. (107), Fig. 50, and let P be the point.

Analysis. If, at the given point, we draw a tangent to the meridian curve of the surface, and a second tangent to the circle of the surface at this point, the plane of these two lines will be the required plane, Art. (108).

Construction. Through P pass a meridian plane; *cp* will be its horizontal trace. Revolve this plane about the axis of the surface until it is parallel to V. R will be the revolved position of the point of contact. At *r'* draw *r'x'* tangent to *c'n'd'm'*. It will be the vertical projection of the revolved position of a tangent to the meridian curve at R. When the plane is revolved to its primitive position, the point, of which *y'* is the vertical projection, remains fixed, and *y'p'* will be the vertical projection of the tangent, and *cp* its horizontal projection. It pierces H at *x*, one point of the horizontal trace of the required plane. *pr* is the horizontal, and *p'r'* the vertical projection of an arc of the circle of the surface containing P. At *p* draw *pu* perpendicular to *cp*. It will be the horizontal projection of the tangent to the circle at P, and *p'u'* is its vertical projection This pierces V at *u'*, a point of the vertical trace. Through *x* draw *x*T, parallel to *pu* and through T draw T*u'*; *x*T*u'* will be the required plane.

147. *Second method* for the same problem.

Analysis. If the tangent to the meridian curve at P revolve about the axis of the surface, it will generate a right cone tangent to the surface in a circumference containing the given point, Art. (121). If, at this point, a tangent plane be drawn to the cone, it will be tangent also to the surface.

Construction. Draw the tangent at P, as in the preceding article. It pierces H at x, and this point, during the revolution, describes the base of the cone whose vertex is at (cy'). The element of the cone through P pierces H at x, and xTt' is the required plane.

148. By the same methods a tangent plane may be passed to any surface of revolution at a given point.

Since the tangent plane and horizontal plane are both perpendicular to the meridian plane through the point of contact, their intersection, which is the horizontal trace, will be perpendicular to the meridian plane and to its horizontal trace.

While at a given point on a double curved surface only one tangent plane can be passed, it may be proved as in Art. (142), that from a point without the surface, an infinite number of such planes can be passed.

149. PROBLEM 35. *To pass a plane through a given right line and tangent to a sphere.*

Let the ground line pass through the centre of the sphere, and let C, Fig. 60, be the centre, and def the circle, cut from the sphere by the horizontal plane, and let MN be the given line.

Analysis. If we take any point of the given line as the vertex of a right cone, tangent to the sphere, and pass a plane through the line tangent to this cone, it will be tangent also to the sphere

Construction. Take m as the vertex of the auxiliary cone, and draw the tangents md and me, and the right line mC. The tangents will be the two rectilinear elements of the cone in the hori zontal plane, and *m*C will be its axis. Since the line of contact of the two surfaces is a circumference, whose plane is perpendicular to *m*C, Art. (121), this plane will be perpendicular to H, and *de* will be its horizontal trace. This circumference may be regarded as the base of the cone; and, if we find the point in which its plane is pierced by MN, and draw from this point a tangent to the base, it will be a line of the required plane, Art. (112). O is this point, Art. (15). If the plane of the base be revolved about *de* as an axis, until it coincides with H, O will fall at *o''*, *o''o* being equal to *o'q*, Art. (17). The circle of contact will take the position *dp''e*, *de* being its diameter. Draw *o''p''*; it will be the revolved position of the tangent. It pierces H at *t*, one point of the horizontal trace; MN, pierces H at *m*; hence, *tm* is the required horizontal trace. MN pierces V at *n'*, a point of the vertical trace. A second point, *t'*, may be determined as in Art. (123), and *n't'* is the vertical trace. When the plane of the circle of contact is in its true position, *p''* is horizontally projected at *p*, and vertically at *p'*, *p'r* being equal to *pp''*; hence, CP will be the radius passing through P. Since the tangent plane must be perpendicular to this radius, C*p* and C*p'* must be respectively perpendicular to *tm* and *t'n'*.

Since a second tangent can be drawn from O to the base of the cone, another tangent plane may be constructed.

150. *Second method* for the same problem.

Let the sphere and the right line MN be given as in Fig. 61.

Analysis. If any two points of the given line be taken, each as the vertex of a cone tangent to the sphere, each cone will be tangent in the circumference of a circle, and the planes of these circles will intersect in a right line, which will pierce the surface of the sphere at the intersection of the circumferences, and these points will be common to the three surfaces. If, through either of these points and the given line, we pass a plane, it will be tangent to both cones and to the sphere.

Construction. Take the points m and n' as the vertices of the two auxiliary cones; de is the horizontal projection of the circle of contact of the first cone with the sphere, and fg is the vertical projection of the circle of contact of the second cone and sphere, Art. (15). The planes of these two circles intersect in a right line, of which de is the horizontal, and fg the vertical projection, Art. (15); and this line pierces H at o. Now revolve the circle of which de is the diameter, about de as an axis, until it coincides with H. It takes the position $dsex$. Any point of the line (de, fg), as R, falls at r'', and or'' will be the revolved position of the line of intersection of the two planes, and p'' and q'' will be the revolved positions of the two points in which it pierces the surface of the sphere. After the counter-revolution, these points are horizontally, projected at p and q, and vertically at p and q'. CP is the radius of the sphere at P. The traces tT, and $t'n'$, of the plane tangent at P, may now be drawn as in the preceding article; or by drawing mt perpendicular to Cp, and $n't'$ perpendicular to Cp', Art. (43). A second tangent plane at Q may be determined in the same way.

151. *Third method* for the same problem.

Let C, Fig. 62, be the centre, and def the horizontal, and $d'h'f'$ the vertical projection of the sphere, and let MN be the given line.

Analysis. Conceive the sphere to be circumscribed by a cylinder of revolution, whose axis is parallel to the given line. The line of contact will be the circumference of a great circle perpendicular to the axis and given line, Art. (121). A plane through the right line tangent to this cylinder will be tangent also to the sphere. The plane of the circle of contact will intersect the given line in a point, and the required tangent plane in a right line drawn from this point tangent to the circle. The plane of this tangent and the given line will be the required plane. Without constructing the cylinder, we have then simply to pass a plane through the centre of the sphere perpendicular to the given line, and from the point in which it intersects the line, to draw a tangent to the circle cut from the sphere by the same plane, and pass a plane through this tangent and the given line.

Construction. Through C draw the two lines CP and CQ, as in Art. (46). The plane of these two lines is perpendicular to MN, and intersects it in O, Art. (41). The horizontal trace of this plane may now be determined as in Art. (46), and the plane revolved about this trace until it coincides with H, and the revolved position of the point and circle be found, and the tangent drawn.

Otherwise thus: Revolve this plane about CP, until it becomes parallel to H. The point O, in its revolved position, will be horizontally projected at r, and the circle will be horizontally projected in def, Art. (62). From r draw the two tangents rk and rl. They will be the horizontal projections of the revolved positions of the two tangents to the great circle, cut from the sphere by the perpendicular plane, and k and l will be the horizontal projections of the revolved positions of the two points of contact of the required tangent plane. After the counter-revolution, R will be horizontally projected at o; and since X remains fixed, oy will be the horizontal, and $o'y'$ the vertical projection of the first tangent, and Y its point of contact.

Since the second tangent does not intersect PC within the limits of the drawing, draw the auxiliary line gp, and find the true horizontal projection of the point which in revolved position is horizontally projected in g, as in Art. (37). It will be at g'', and og'' will be the horizontal projection of the second tangent, and z the horizontal projection of the second point of contact. S is the point in which this tangent intersects CQ, $o's'$ is its vertical projection, and Z is the second point of contact. A plane through MN, and each of these points will be tangent to the sphere.

152. PROBLEM 36. *To pass a plane through a given right line and tangent to any surface of revolution.*

Let the horizontal plane be taken perpendicular to the axis, of which c, Fig. 63, is the horizontal, and $c'd'$ the vertical projection. Let poq be the intersection of the surface by the horizontal plane, $p'd'q'$ the vertical projection of the meridian curve parallel to the vertical plane, and MN the given line.

Analysis. If this line revolve about the axis of the surface, it will generate a hyperboloid of revolution of one nappe, Art. (99), having the same axis as the given surface. If we now conceive the plane to be passed tangent to the surface, it will also be tangent to the hyperboloid at a point of the given right line, Art. (114); and since the meridian plane through the point of contact on each surface must be perpendicular to the common tangent plane, Art. (115), these meridian planes must form one and the same plane. This plane will cut from the given surface a meridian curve, from the hyperboloid an hyperbola, Art. (104), and from the tangent plane a right line tangent to these curves at the required points of contact, Art. (108). The plane of this tangent and the given line will therefore be the required plane.

Construction. Construct the hyperboloid as in Art. (99). yx will be the horizontal, and $x'y'x''$ the vertical projection of one branch of the meridian curve parallel to V. If the meridian plane through the required points of contact be revolved about the common axis until it becomes parallel to V, the corresponding meridian curves will be projected, one into the curve $p'd'q'$, and the other into the hyperbola $x'y'x''$. Tangent to these curves draw $x'r'$; X will be the revolved position of the point of contact on the hyperbola, and R that on the meridian curve of the given surface. When the meridian plane is revolved to its true position, X will be horizontally projected in mn at s; sc will be the horizontal trace of the meridian plane, and u will be the horizontal, and u' the vertical projection of the point of contact on the given surface. A plane through this point and MN will be the tangent plane.

153. In general, through a given right line a limited number of planes only can be passed tangent to a double curved surface For let the surface be intersected by a system of planes parallel to the given line, and tangents be drawn to the sections also parallel to the line. These will form a cylinder tangent to the surface. Any plane through the right line tangent to this cylinder will be tangent to the surface; and the number of tangent planes will be determined by the number of tangents which can be drawn from a point of the given line to a section made by a plane through this point.

POINTS IN WHICH SURFACES ARE PIERCED BY LINES.

154. The points in which a right line pierces a surface are easily found by passing through the line any plane intersecting the surface. It will cut from the surface a line, which will intersect the given line in the required points. This auxiliary plane should be so chosen as to intersect the surface in the simplest line possible.

If the given surface be a cylinder, a plane through the right line parallel to the rectilinear elements should be used. It will intersect the cylinder in one or more rectilinear elements, which will intersect the given line in the required points.

If the given surface be a cone, the auxiliary plane should pass through the vertex.

If a sphere, the auxiliary plane should pass through the centre, as the line of intersection will be a circumference with a radius equal to that of the sphere.

155. If the given line be a curve of single curvature, its plane will intersect the surface, if at all, in a line which will contain the required points. If the plane does not intersect the surface, or if the line of intersection does not intersect, or is not tangent to the

given curve, there will, of course, be no points common to the curve and surface.

If the given line be a curve of double curvature, a cylinder or cone may be passed through it intersecting the given surface. If the line of intersection intersects the given line, the points thus determined will be the required points.

These problems will be easily solved when the more general problem of finding the intersection of surfaces has been discussed.

INTERSECTION OF SURFACES BY PLANES. DEVELOPMENT OF SINGLE CURVED SURFACES.

156. The solution of the problem of the intersection of surfaces consists in finding two lines, one on each surface, which intersect. The points of intersection will be points in both surfaces, and therefore points of their line of intersection.

157. *To find the intersection of a plane with a surface*, we intersect the plane and surface by *a system of auxiliary planes.* Each plane will cut from the given plane a right line, and from the surface a line, *the intersection of which will be points of the required line.* The system of auxiliary planes should be so chosen as to cut from the surface the simplest lines; a rectilinear element, if possible, or the circumference of a circle, &c.

The curve of intersection may be drawn with greater accuracy by determining at each of the points thus found a tangent to the curve, and then drawing the projections of the curve tangent to the projections of these tangents, Art. (65).

This tangent must lie in the intersecting plane, that is, in the plane of the curve, Art. (64). It must also lie in the tangent plane to the surface at the given point, Art. (108). Hence, *if we pass a plane tangent to the given surface at the given point, and determine its intersection with the intersecting plane; this will be the required tangent.*

158. Since the tangent plane to a single curved surface in general contains two consecutive rectilinear elements, Art. (111) it will contain the elementary portion of the surface generated by the generatrix in moving from the first element to the second. Now if the surface be rolled over until the consecutive element following the second comes into the tangent plane, the portion of the plane limited by the first and third elements will equal the portion of the surface limited by the same elements. If we continue to roll the surface on the tangent plane until any following element comes into it, the portion of the plane included between this element and the first will be equal to the portion of the surface limited by the same elements. Therefore if a single curved surface be rolled over on any one of its tangent planes until each of its rectilinear elements has come into this plane, *the portion of the plane thus touched by the surface, and limited by the extreme elements*, will be a plane surface equal to the given surface, and is *the development of the surface.*

As a tangent plane to a warped surface cannot contain two consecutive rectilinear elements, Art. (113), the elementary surface limited by these two elements cannot be brought into a plane without breaking the continuity of the surface. *A warped surface*, therefore, *cannot be developed.*

Neither can a double curved surface be developed; as any elementary portion of the surface will be limited by two curves, and cannot be brought into a plane without breaking the continuitv of the surface.

159. In order to determine the position of the different rectilinear elements of a single curved surface, as they come into the tangent plane, or *plane of development*, it will always be necessary

to find some curve upon the surface which will develop into a right line, or circle, or some simple known curve, upon which the rectified distances between these elements can be laid off.

160. Problem 37. *To find the intersection of a right cylinder with a circular base by a plane.*

Let *mlo*, Fig. (64), be the base of the cylinder, and *c* the horizontal, and *c'd'* the vertical projection of the axis; then *mlo* will be the horizontal, and *s'm''o''u'* the vertical projection of the cylinder, Art. (75). Let *t*T*t'* be the intersecting plane.

Analysis. Intersect the cylinder and plane by a system of auxiliary planes parallel to the axis, and also to the horizontal trace of the given plane. These will cut from the cylinder rectilinear elements, and from the plane right lines parallel to the horizontal trace. The intersection of these lines will be points of the required curve, Art. (157).

This curve, as also the intersection of any cylinder or cone with a circular or an elliptical base, by a plane cutting all the rectilinear elements, is an ellipse. *Analyt. Geo.*, Arts. (82 and 200).

Construction. Draw *xy* parallel to *t*T, it will be the horizontal trace of one of the auxiliary planes; *yy'* is its vertical trace. It intersects the cylinder in two elements, one of which pierces H at *x*, and the other at *z*, and *x''x'* and *z''z'* are their vertical projections. It intersects the plane in the right line XY, Art. (38), and X and Z, the intersections of this line with the two elements, are points of the curve. In the same way any number of points may be found.

If a point of the curve on any particular element be required, we have simply to pass an auxiliary plane through this element. Thus, to construct the point on (*o*, *o''u'*), draw the trace *ow*, and construct as above the point O.

Since this point lies on the curve, and also on the extreme element, and since no point of the curve can be vertically projected outside of $o''u'$, the vertical projection of the curve must be tangent to $o''u'$ at o'. Or, the reason may be given thus, in many like cases: If a tangent be drawn to the curve at O, it will lie in the tangent plane to the surface at O, Art. (108); and since this tangent plane is perpendicular to the vertical plane, the tangent will be vertically projected into its trace $o''u'$, which must therefore be tangent to $m'o'l'$ at o', Art. (65). Also, $m''s'$ must be tangent to $m'o'l'$ at m'.

If a plane be passed through the axis perpendicular to the intersecting plane, it will evidently cut from the plane a right line, which will bisect all the chords of the curve perpendicular to it, and this line will be the transverse axis of the ellipse. lc is the horizontal trace of such a plane. It cuts the given plane in KG, Art. (38), and the cylinder in two elements horizontally projected at k and l, and KL is the transverse axis, and K and L the vertices of the ellipse. K is the lowest, and L the highest point of the curve.

Since the curve lies on the surface of the cylinder, its horizontal projection will be in the base mlo.

To draw a tangent to the curve at any point, as X; draw xr tangent to xol at x; it is the horizontal trace of a tangent plane to the cylinder at X, Art. (123). This plane intersects tTt' in a right line, which pierces H at r; and since X is also a point of the intersection, RX will be the required tangent, Art. (157). If a tangent be drawn at each of the points determined as above, the projections of the curve can be drawn, with great accuracy, tangent to the projections of these tangents, at the projections of the points of tangency.

The part of the curve MXO, between the points M and O, lies in front of the extreme elements (m, $m''s'$) and (o, $o''u'$), and is seen, and therefore $m'x'o'$ is drawn full, and $m'l'o'$ broken.

To represent the curve in its true dimensions let its plane be revolved about tT until it coincides with H. The revolved position of each point, as X, at x''', will be determined as in Art. (17) k''' and l''' will be the revolved position of the vertices, rx''' will be the revolved position of the tangent, and $l'''x'''k'''$ the ellipse in its true size.

161. Problem 38. *To develop a right cylinder with a circular base, and trace upon the development the curve of intersection of th cylinder by an oblique plane.*

Let the cylinder and curve be given as in the preceding problem, and let the plane of development be the tangent plane at L.

Analysis. Since the plane of the base is perpendicular to the element of contact of the tangent plane, it is evident that as the cylinder is rolled out on this plane, this base will develop into a right line, on which we can lay off the rectified distances between the several elements, and then draw them each parallel to the element of contact.

Points of the development of the curve are found by laying off on the development of each element, from the point where it meets the rectified base, a distance equal to the distance of the point from the base.

Construction. The plane of development being coincident with the plane of the paper, let *l*L, Fig. 65, be the element of contact. Draw *ll* perpendicular to *l*L, and lay off *ll* equal to the rectified circumference *lw*--*k*--*l*, Fig. 64. It will be the development of the base. Lay off *lw* equal to the arc *lw*, and draw *w*W parallel to *l*L. It is the development of the element which pierces H at *w*. Likewise for each of the elements lay off *wz*, *zm*, &c., equal respectively to the rectified arcs *wz*, *zm*, &c., in Fig. 64, and draw *z*Z, *m*M, &c. The portion of the plane included between *l*L and *l*L will be the development of the cylinder.

On *l*L lay off $l\text{L} = l''l'$, L will be one point of the developed curve; on *w*W lay off $w\text{W} = w''w'$, W will be a second point; and thus each point may be determined, and L--K--L will be the developed curve. The line *rx*, the *sub-tangent*, will take the position *rx* on the developed base, and X*r* will be the tangent at X in the plane of development. This must be tangent at X, since the tangent after development must contain the same two consecutive points which it contains in space, and therefore be tangent to the development of the curve, Art. (64).

162. PROBLEM 39. *To find the intersection of an oblique cylinder by a plane.*

Let the cylinder be given as in Fig. 66, and let $t\text{T}t'$, perpendicular to the rectilinear elements, be the intersecting plane.

Analysis. Intersect the cylinder and plane by a system of auxiliary planes parallel to the rectilinear elements, and perpendicular to the horizontal plane. These planes will each intersect the cylinder in two rectilinear elements, and the plane in a right line, the intersection of which will be points of the curve

Construction. Let eq, parallel to li, be the horizontal trace of an auxiliary plane; qq' will be its vertical trace. It intersects the cylinder in two elements, one of which pierces H at r, the other at s, and these are vertically projected in $r'g'$ and $s'h'$, and horizontally in eq. It intersects $t\text{T}t'$ in a right line, which pierces H at e, and is vertically projected in $e'q'$, and horizontally in eq. These lines intersect in Z and Y, Art. (23), points of the curve. In the same way any number of points may be determined.

The auxiliary planes, being parallel, must intersect $t\text{T}t'$ in parallel lines, the vertical projections of which will be parallel to $e'q'$.

By the plane, whose horizontal trace is mn, the points U and X are determined. The vertical projection of the curve must be tangent to $m'n'$ at u'. The points in which the horizontal projection is tangent to li and kf, are determined by using these lines as the traces of auxiliary planes.

To draw a tangent to the curve at any point, as X, pass a plane tangent to the cylinder at X; av is its horizontal trace, and vx the horizontal, and $v'x'$ the vertical projection of the tangent.

A sufficient number of points and tangents being thus determined, the projections of the curve can be drawn with accuracy

The part *cyd* is full, being the horizontal projection of that part of the curve which lies above the extreme elements LI and KF. For a similar reason, *u'x'w'* is full.

To show the curve in its true dimensions, revolve the plane about *t*T until it coincides with H. The revolved position of each point may be found as in Art. (17), and $c''y''d''z''$ will be the curve in its true size.

The section thus made is *a right section.*

163. *If it be required to develop the cylinder* on a tangent plane along any element, as KF, we first make a right section as above. We know this will develop into a right line perpendicular to KF. On this we lay off the rectified arcs of the section included between the several elements, and then draw these elements parallel to KF.

The developed base, *or any curve on the surface*, may be traced on the plane of development by laying off on each element, from the developed position of the point where it intersects the right section, the distance from this point to the point where the element intersects the base or curve. A line through the extremities of these distances will be the required development.

164. PROBLEM 40. *To find the intersection of a right cone with a circular base by a plane.*

Let the cone be given as in Fig. 67, and let *t*T*t'* be the given plane, the vertical plane being assumed perpendicular to it.

Analysis. Intersect the cone by a system of planes through the vertex and perpendicular to the vertical plane. The elements cut from the cone by each plane, Art. (157), will intersect the right line cut from the given plane in points of the required curve.

Construction. Let *lk* be the horizontal trace of an auxiliary plane, *ks'* will be its vertical trace. It intersects the cone in two elements, one of which pierces H in l, and the other in i, horizontally projected in ls and is respectively. It intersects the given plane in a right line perpendicular to V, vertically projected at x', and horizontally in xv; hence x and v are the horizontal projections of two points of the curve, both vertically projected at x'. In the same way any number of points can be found, as Y (wy'), &c.

The plane whose horizontal trace is mo, perpendicular to tTt', intersects it in a right line vertically projected in Tt', which evidently bisects all chords of the curve perpendicular to it, and is therefore an axis of the curve. This plane cuts from the cone the elements SM and SO, which are intersected by the axis in the points Z and U, which are the vertices.

To draw a tangent to the curve at any point, as X, pass a plane tangent to the cone at X, lr is its horizontal trace. It intersects tTt' in (rx, Tx'), which is therefore the required tangent, Art. (157). The horizontal projection, $uxzv$, can now be drawn. $u'z'$ is its vertical projection.

To represent the curve in its true dimensions, we may revolve it about tT until it coincides with H, or about Tt' until it coincides with V, and determine it as in Art. (17). Otherwise, thus: Revolve it about UZ until its plane becomes parallel to V, it will then be vertically projected in its true dimensions, Art. (62). The points U and Z being in the axis, will be projected at u' and z' respectively. X will be vertically projected at x'', $x'x''$ being equal to px. Y at y'', $y'y''$ being equal to qy. (wy') at w'', &c., and $u'x''z'x'''$ will be the curve in its true size.

165. If a right cone, with a circular base, be intersected by a plane, as in Fig. 67, making *a less angle* with the plane of the base than the elements do, the curve of intersection is *an ellipse*, *Analyt. Geo*, Art. (82.) If it make *the same angle*, or is parallel to one of the elements, the curve is *a parabola*. If it make a

greater angle, the curve is *an hyperbola.* Hence these three curves are known by the general name, *conic sections.*

166. Problem 41. *To develop a right cone with a circular base.*

Let the cone and its intersection by an oblique plane be given as in the preceding problem, Fig. 67, and let the plane of development be the tangent plane along the element MS; the half of the cone in front being rolled to the left, and the other half to the right.

Analysis. Since the base of the cone is everywhere equally distant from the vertex, as the cone is rolled out, each point of this base will be in the circumference of a circle described with the vertex as a centre, and a radius equal to the distance from the vertex to any point of the base. By laying off on this circumference the rectified arc of the base, contained between any two elements, and drawing right lines from the extremities to the vertex, we have, in the plane of development, the position of these elements. Laying off on the proper elements the distances from the vertex to the different points of the curve of intersection, and tracing a curve through the extremities, we have the development of the curve of intersection.

Construction. With SM = $s'm'$, Figs. 67 and 68, describe the arc OMO. It is the development of the base. Lay off MG = mg, and draw SG; it is the developed position of the element SG. In the same way lay off GL = gl, LO = lo, and draw SL and SO. OSM is the development of one-half the cone. In like manner the other half may be developed.

On SM lay off SZ = $s'z'$. Z will be the position of the point in the plane of development. To obtain the true distance from S to Y, revolve SY about the axis of the cone until it becomes parallel to V, as in Art. (28); $s'e'$ will be its true length. On SG lay off SY = $s'e'$; on SL, SX = $s'd'$; on SM, SU = $s'u'$. Z, Y, X, and U will be the position of these points on the plane of development, and UX - - - U will be the development of the curve of intersection.

Through L draw LR perpendicular to LS, and make it equal to *lr*. RX will be the developed tangent.

167. PROBLEM 42. *To find the intersection of any cone by a plane.*

Let the cone and plane *t*T*t'* be given as in Fig. 69.

Analysis. Intersect the cone by a system of planes through the vertex and perpendicular to the horizontal plane. Each of these planes will intersect the cone in one or more rectilinear elements, and the given plane in a right line, the intersection of which will be points of the curve. Since these auxiliary planes are perpendicular to the horizontal plane, they will intersect in a right line through the vertex, perpendicular to the horizontal plane, and the point in which this line pierces the cutting plane will be a point common to all the right lines cut from this plane.

Construction. Find the point in which the perpendicular to H, through S, pierces *t*T*t'*, as in Art. (42). *v'* is its vertical projection. The vertical projections of the lines cut from *t*T*t'* all pass through this point.

Let *sp* be the horizontal trace of an auxiliary plane. It intersects the cone in the elements SE and SD, and the cutting plane in the right line (*ps*, *p'v'*). This line intersects the elements in R and Y, which are points of the required curve. In the same way any number of points may be found.

To find the point of the curve on any particular element, as SM, we pass an auxiliary plane through this element. *sm* is its horizontal trace, and Z and X the two points on this element. The vertical projection of the curve is tangent to *s'm'* at *z'*, and to *s'o'* at *u'*. The points *q* and *w*, in which the horizontal projection is tangent to *sl* and *sn*, are found by using as auxiliary planes the two planes whose traces are *sl* and *sn*.

To draw a tangent to the curve at any point, as X, pass a plane tangent to the cone at X. *ic* is its horizontal trace, Art. (129). It intersects *t*T*t'* in *c*X, which is therefore the required tangent, Art. (157).

The part of the curve which lies above the two extreme elements SL and SN, is seen, and therefore its projection, *wyzq*, is full. For a similar reason the projection, *z'q'x'u'*, of that part of the curve which lies in front of the two extreme elements, SM and SO, is full.

To show the curve in its true dimensions, revolve the plane about *t*T until it coincides with H, and determine each point, as Q at *q''*, as in Art. (17). Or the position of (*sv'*) may be found at *v''*. Then if the points *c*, *a*, *b*, *p*, &c., be each joined with this point by right lines, we have the revolved positions of the lines cut from the given plane by the auxiliary planes, and the points *y''*, *z''*, *r''*, *x''*, in which these lines are intersected by the perpendiculars to the axis, *yy''*, *zz''*, &c., are points of the revolved position of the curve. *x''c* is the revolved position of the tangent.

168. The intersection of the single curved surface, with a helical directrix, by a plane, may be found by intersecting by a system of auxiliary planes tangent to the projecting cylinder of the helical directrix. These intersect the surface in rectilinear elements, and the plane in right lines, the intersection of which will be points of the required curve.

169. PROBLEM 43. *To find the intersection of any surface of revolution by a plane.*

Let the surface be a hyperboloid of revolution of one nappe given as in Fig. 70, and let *t*T*t'* be the cutting plane.

Analysis. If a meridian plane be passed perpendicular to the cutting plane, it will intersect it in a right line, which will divide the curve symmetrically, and be an axis. If the points in which this line pierces the surface be found, these will be the vertices of the curve. Now intersect by a system of planes perpendicular to the axis of the surface; each plane will cut from the surface a circumference and from the given plane a right line, the intersection of which will be points of the required curve.

Construction Draw cn perpendicular to tT, it will be the horizontal trace of the auxiliary meridian plane. This plane intersects tTt' in the right line NC, Art. (38), and the surface in a meridian curve which is intersected by NC in the two vertices of the required curve; or in its highest and lowest points.

To find these points, revolve the meridian plane about the axis until it becomes parallel to V. The meridian curve will be vertically projected into the hyperbola, which limits the vertical projection of the surface, Art. (62), and the line NC into $d'c'$. The points s' and r' will be the vertical projections of the revolved positions of the vertices. After the counter revolution, these points are horizontally projected at k and l, and vertically at k' and l'.

Let $u'z'$ be the vertical trace of an auxiliary plane perpendicular to the axis. It cuts the surface in a circle horizontally projected in wuz, and the plane in a right line, which piercing V at e' is horizontally projected in ez; hence U and Z are points of the curve. In the same way any number of points may be determined.

The points o' and v', in which the vertical projection of the curve is tangent to the hyperbola which limits the projection of the surface, are the vertical projections of the points in which the line, cut out of the given plane by the meridian plane parallel to the vertical plane, intersects the meridian curve cut out by the same plane.

The points upon any particular circle may be determined by using the plane of this circle as an auxiliary plane. If the curve crosses the circle of the gorge, the points in which it crosses are determined by using the plane of this circle, and, in this case, the horizontal projection of the curve must be tangent to the horizontal projection of the circle of the gorge, at the points x and y.

The method given above for determining the vertices of the

curve is applicable to any surface of revolution. In this particular surface it may be modified thus: Let the line NL revolve about the axis; it will generate a cone of revolution whose base is *dni*, and vertex C. This cone intersects the hyperboloid in two circumferences, Art. (97), and the points in which NL intersects these circumferences will be the points required.

To construct them; through any rectilinear element of the hyperboloid, as MQ, and the vertex of the cone, pass a plane; *qj* will be its horizontal trace, Art. (33). It intersects the cone in two elements horizontally projected in *ci* and *cg*, and the points *a* and *b* are points in the horizontal projections of the circles in which the hyperboloid and cone intersect, and K and L are the required vertices.

To draw a tangent to the curve at Z, pass a plane tangent to the surface at Z, as in Art. (138); its intersection with *t*T*t'* will be the tangent, Art. (157).

The curve may be represented in its true dimensions as in Art. (162).

Let the intersection of an oblique plane with a sphere, an ellipsoid of revolution and paraboloid of revolution, be constructed in accordance with the principles of the preceding problem.

170. To find the intersection of any warped surface with a plane directer, by an oblique plane, intersect by planes parallel to the plane directer. Each will cut from the surface one or more rectilinear elements, and from the plane a right line, the intersection of which will be points of the required curve.

171. *To find the intersection of a helicoid by a plane*, the surface being given as in Art. (92); intersect by a system of auxiliary planes through the axis. These will cut from the surface rectilinear elements, and from the plane right lines, the intersection of which will be points of the required curve.

Let the construction be made, and the curve and its tangent represented in true dimensions.

INTERSECTION OF CURVED SURFACES.

172. *To find the intersection of any two curved surfaces*, we intersect them by *a system of auxiliary surfaces.* Each auxiliary surface will cut from the given surfaces lines *the intersection of which will be points of the required line.* Art. (156).

The system of auxiliary surfaces should be so chosen as to cut from the given surfaces the simplest lines, rectilinear elements if possible, or the circumferences of circles, &c.

To draw a tangent to the curve of intersection at any point; pass a plane tangent to each surface at this point. *The intersection of these two planes will be the required tangent*, since it must lie in each of the tangent planes. Art. (108).

In constructing this curve of intersection, great care should be taken to determine those points in which its projections are tangent to the limiting lines of the projections of the surfaces; and also those points in which the curve itself is tangent to other lines of either surface, as these points aid much in drawing the curve with accuracy.

173. Problem 44. *To find the intersection of a cylinder and cone.*

Let the surfaces be given as in Fig. 71; the base of the cylinder *bai* being in the horizontal plane, and its rectilinear elements parallel to the vertical plane; the base of the cone $m'l'o'$, in the vertical plane, and S its vertex.

Analysis. Intersect the two surfaces by a system of auxiliary planes *passing through the vertex of the cone and parallel to the rectilinear elements of the cylinder.* Each plane will intersect each of the surfaces in two rectilinear elements, the intersection of which will be points of the required curve. These planes will intersect in a right line passing through the vertex of the cone and parallel to the rectilinear elements of the cylinder, and the point, in which this line pierces the horizontal plane, will be a point common to the horizontal traces of all the auxiliary planes.

Construction. Through S draw ST parallel to BE. It pierces H in *t*. Through *t* draw any right line, as *th*; it may be taken as the horizontal trace of an auxiliary plane, the vertical trace of which is *hg'* parallel to *b'e'*. This plane intersects the cylinder in two elements which pierce H at *g* and *i*, and are horizontally projected in *gg''* and *ii''*. The same plane intersects the cone in two elements which pierce V in *m'* and *n'*, and are horizontally projected in *ms* and *ns*. These elements intersect in the points X, Y Z and U, which are points of the required curve. In the same way any number of points may be determined.

The horizontal projection of the required curve is tangent to *ms* at the points *x* and *y*; since *ms* is the horizontal projection of one of the extreme elements of the cone. The points of tangency on *sl* may be determined by using an auxiliary plane, which shall contain the element SL.

The points of tangency on *dw* are obtained in the same way, by using the plane whose horizontal trace is *td*.

The points in which the vertical projection of the curve is tangent to the vertical projections of the extreme elements of both cone and cylinder will be determined, by using as auxiliary planes those which contain these elements.

A tangent to the curve at any of the points, thus determined, may be constructed by finding the intersection of two planes, one tangent to the cylinder and the other to the cone, at this point. Art. (172).

The plane, of which *tc* is the horizontal trace, is tangent to the cylinder along the element CQ, and intersects the cone in the two elements SP and SQ. If at P a plane be passed tangent to the cone, it will be tangent along SP, which is also its intersection with the tangent plane to the cylinder. SP is then tangent to the curve at P, and for a similar reason SQ is tangent at Q. Hence the two projections sp and sq are tangent to xpy--qu at p and q respectively; and the vertical projections of the same elements will be tangent at p' and q'.

The plane of which tc' is the horizontal trace, is tangent to the cone along the element SK, and intersects the cylinder in two elements, the horizontal projections of which are tangent to xp--qv at k and v.

The projections of the curve can now be drawn with great accuracy. The horizontal projection of that part which lies above the two elements SM and SL, and also above the element DW, is drawn full. Likewise the vertical projection of that part which lies in front of the element BE, and also in front of the two extreme elements of the cone is full.

If two auxiliary planes be passed tangent to the cylinder, and both intersect the cone, it is evident that the cylinder will penetrate the cone, so as to form two distinct curves of intersection. If one intersects the cone and the other does not, a portion only of the cylinder enters the cone, and there will be a continuous curve of intersection, as in the figure.

If neither of these planes intersects the cone, and the cone lies between them, the cone will penetrate the cylinder, making two distinct curves.

If both planes are tangent to the cone, all the rectilinear elements of both surfaces will be cut.

174. *The intersection of two cylinders* may be found, by passing a plane through a rectilinear element of one cylinder parallel to the rectilinear elements of the other, and then intersecting the cylinders by a system of planes parallel to this plane. The horizontal traces of these planes will be parallel, and the construction will be in all respects similar to that of the preceding problem.

The intersection of two cones may also be found by using a system of planes, *through the vertices of both cones.* The right line, which joins these vertices, will lie in all of these planes, and pierce the horizontal plane in a point common to all the horizontal traces.

We may ascertain whether the cylinders or cones intersect in two distinct curves, or only one, in the same manner as in Art. (173), by passing auxiliary planes tangent to either cylinder or cone.

175. Problem 45. *To find the intersection of a cone and helicoid.*

Let *mnl*, Fig. 72, be the base of the cone, S its vertex; and let *prq* be the horizontal, and *p′o″q′* the vertical projection of the helical directrix, (*s*, *o′s′*) being the axis, and the rectilinear generatrix being parallel to the horizontal plane, Art. (92).

Analysis. Intersect the surfaces by a system of auxiliary planes through the axis of the helicoid. These planes cut from both surfaces rectilinear elements, which intersect in points of the required curve.

Construction. Draw *sk* as the horizontal trace of an auxiliary plane. This plane intersects the cone in the element SK, and the helicoid in an element of which *sg* is the horizontal and *h′g′* the vertical projection. These intersect in X, a point of the required curve. In the same way, Y, Z, and other points, are determined. At the points *m′* and *z′*, the vertical projection of the curve is tangent to *s′m′* and *s′l′*, Art. (160).

To draw a tangent to the curve at X; pass a plane tangent to each of the surfaces at X, Arts. (129) and (139): *ku* is the horizontal trace of the plane tangent to the cone, and *tw*, parallel to *sg*, that of the plane tangent to the helicoid, and their intersection UX, is the required tangent line.

176. Problem 46. *To find the intersection of a cylinder and hemisphere.*

Let *mnl* Fig. 73, be the base of the cylinder and MX a rectilinear element. Let *ced* be the horizontal and *c'f'd'* the vertical projection of the hemisphere.

Analysis. Intersect the surfaces by a system of auxiliary planes parallel to the rectilinear elements of the cylinder and perpendicular to the horizontal plane. Each plane will cut from the cylinder two rectilinear elements, and from the sphere a semicircumference the intersection of which will be points of the required curve.

Construction. Take *pn* as the trace of an auxiliary plane. It cuts from the cylinder the two elements PY and NZ, and from the hemisphere a semicircumference of which *i* is the centre and *ig* the radius. Revolve this plane about *pn* until it coincides with H. *pi''* will be the revolved position of PY, Art. (28), and *nz''* parallel to *pi''* the revolved position of NZ; *grz''* the revolved position of the semicircumference; *y''* and *z''* the revolved position of the required points, and Y and Z the points in their true position.

In the same way any number of points may be determined.

The points of tangency *x* and *w* are found by using *mx* and *lw* as the traces of auxiliary planes, and *r'* and *u'*, by using *or* and *su*.

A tangent at any point may be constructed by finding the intersection of two planes tangent to the cylinder and sphere as in Arts. (123) and (145).

The tangents at Z and Y are evidently parallel to H.

177. Problem 47. *To find the intersection of a cone and hemisphere.*

Let *mlo*, Fig. 74, be the base of the cone, and S its vertex, at the centre of the sphere; *abc* being the horizontal and *a'd'c'* the vertical projection of the hemisphere.

Analysis. Intersect the surfaces by a system of planes passing through the vertex and perpendicular to the horizontal plane. Each plane will cut from the cone two rectilinear elements, and from the hemisphere a semicircumference, the intersection of which will be points of the required curve.

Construction. Take *sp* as the horizontal trace of one of the auxiliary planes. It intersects the cone in the two elements SP and SQ, and the hemisphere in a semicircle whose centre is at S. Revolve this plane about the horizontal projecting line of S, until it becomes parallel to V. The element SP will be vertically projected in $s'p'''$, SQ in $s'q'''$, the semicircle in $a'd'c'$, and x'' and y'' will be the vertical projections of the revolved positions of the points of intersection. In the counter revolution these points describe the arcs of horizontal circles and in their true position will be vertically projected at x' and y', and horizontally at x and y.

In the same way any number of points may be found.

The points of tangency u' and z' are found by using auxiliary planes which cut out the extreme elements SM and SO.

The points in which the vertical projection of the curve is tangent to the semicircle $a'd'c'$, are found by using the auxiliary plane whose trace is st.

To draw a tangent to the curve at any point, we pass a plane tangent to the sphere at this point as in Art. (145) and also one tangent to the cone at the same point as in Art (129), and determine their intersection.

178. PROBLEM 48. *To develop an oblique cone with any base.*

Let the cone be given as in the preceding problem, Fig. 74, and let it be developed on the plane tangent along the element SP.

Analysis. If the cone be intersected by a sphere having its centre at the vertex, all the points of the curve of intersection will be at a distance from the vertex equal to the radius of the sphere; hence, when the cone is developed, this curve will develop into the arc of a circle having its centre at the position of the vertex, and its radius equal to that of the sphere. On this we can lay off the rectified arcs of the curve of intersection, included between the several rectilinear elements, Art. (159), and then draw these elements to the position of the vertex.

The developed base, *or any curve on the surface*, may be traced on the plane of development, by laying off on each element, from the vertex, the distance from the vertex to the point where the element intersects the base or curve. A line through the extremities of these distances will be the required development.

Construction. Find as in the preceding problem the curve XUY---. With S, Fig. 75, as a centre, and *sa* as a radius, describe the arc XUR. It is the indefinite development of the intersection of the sphere and cone. Draw SX for the position of the element SX.

To find the distance between any two points measured on the curve XUY---, we first develop its horizontal projecting cylinder on a plane tangent to it at X, as in Art. (161). X'U'R' Fig. *a*, is the development of the curve. On XUR lay off XU=X'U', UR=U'R', &c, and draw SU, SR, &c. These will be the positions of the elements on the plane of development. On these lay off SP=$s'p'''$, SM=$s'm'''$, &c., and join the points PM, &c., and we have the development of the base of the cone.

179. PROBLEM 49. *To find the intersection of two surfaces of revolution, whose axes are in the same plane.*

First, let the axes intersect and let one of the surfaces be an ellipsoid of revolution and the other a paraboloid; and let the horizontal plane be taken perpendicular to the axis of the ellipsoid and the vertical plane parallel to the axes; (c,$c'd'$), Fig. 76, being the axis of the ellipsoid, and (cl, $s'l'$) that of the paraboloid. Let the ellipsoid be represented as in Art. (107) and let $z'f'r'$ be the vertical projection of the paraboloid.

Analysis. Intersect the two surfaces by a system of auxiliary spheres having their centres at the point of intersection of the axes. Each sphere will intersect each surface in the circumference of a circle perpendicular to its axis, Art. (97), and the points of intersection of these circumferences will be points of the required curve.

Construction. With s' as a centre and any radius, $s'q'$, describe the circle $q'p'r'$; it will be the vertical projection of an auxiliary sphere. This sphere intersects the ellipsoid in a circumference vertically projected in $p'q'$, and horizontally in pxq. It intersects the paraboloid in a circumference vertically projected in $r'v'$. These circumferences intersect in two points vertically projected at x', and horizontally at x and x''.

In the same way any number of points may be found.

The points on the greatest circle of the ellipsoid are found by using $s'n'$ as a radius. These points are horizontally projected at u and u'', points of tangency of xzx'' with nom.

The points W and Z are the points in which the meridian curves parallel to V intersect, and are points of the required curve.

Each point of the curve $z'x'w'$ is the vertical projection of two points of the curve of intersection, one in front and the other behind the plane of the axes.

A tangent may be drawn to the curve at any point as X, as in Art. (172). Otherwise thus: If to each surface a normal line be drawn at X, the plane of these two normals will be perpendicular to the tangent plane to each surface at X, and therefore perpendicular to their intersection, which is the required tangent line. Art. (172). Hence if through X a right line be drawn perpendicular to this normal plane, it will be the required tangent.

Since the meridian plane to a surface of revolution is normal to the surface, Art. (115), the normal to each surface at X must lie in the meridian plane of the surface and therefore intersect the axis.

To construct the normal to the ellipsoid, revolve the meridian curve through X, about the axis of the surface, until it becomes parallel to V; it will be projected into $c'n'd'$ and the point X will be vertically projected at q'. Perpendicular to the tangent at q' draw $q'k'$. It will be the vertical projection of the normal in its revolved position. After the counter-revolution, K remaining fixed, $k'x'$ will be the vertical, and cx the horizontal projection of the normal.

In the same way the normal LX to the paraboloid may be constructed.

The line $k'l'$ is the vertical projection of the intersection of the plane of these two normals with the plane of the axes, and is parallel to the vertical trace of the first plane; hence $x't'$ perpendicular to this line is the vertical projection of the required tangent, xi is the horizontal projection of the intersection of the normal plane with the plane of the circle PXQ, and this is parallel to the horizontal trace of the normal plane; hence xt perpendicular to this is the horizontal projection of the required tangent.

Second. If the axes of the two surfaces are parallel, the construction is more simple, as the auxiliary spheres become planes perpendicular to the axes, or parallel to the horizontal plane. Let the construction be made in this case.

PRACTICAL PROBLEMS.

180. In the preceding articles we have all the elementary principles and rules, relating to the orthographic projection. The student who has thoroughly mastered them will have no difficulty in their application.

Let this application now be made to the solution of the following simple problems.

181. PROBLEM 50. *Having given two of the faces of a triedral angle, and the diedral angle opposite one of them, to construct the triedral angle.*

Let dsf and fse'', Fig. 77, be the two given faces and A the given angle opposite fse'', dsf being in the horizontal plane, and the vertical plane perpendicular to the edge sf.

Construct, as in Art. (54), de' the vertical trace of a plane whose horizontal trace is sd and making with H the angle A; sde' will be the true position of the required face. Revolve fse'' about sf until the point e'' comes into de' at e'. This must be the point in which the third edge in true position pierces V. Join $e'f$; it will be the vertical trace of the plane of the face fse'', in true position, and SE will be the third edge.

Revolve *sde'* about *sd* until it coincides with H; *e'* falls at *e'''* Art. (17), and *dse'''* is the true size of the third or required face.

e'fd is the diedral angle opposite the face *dse'*; and *prq*, determined as in Art. (52), is the diedral angle opposite *dsf*.

182. PROBLEM 51. *Having given two diedral angles formed by the faces of a triedral angle, and the face opposite one of them, to construct the angle.*

Let A and B, Fig. 78, be the two diedral angles, and *dse'''* the face opposite B.

Make *e'df* = A. Revolve *dse'''* about *ds* until *e'''* comes into *de'* at *e'*. This will be the point where the edge *se'''*, in true position, pierces V, and SE will be this edge.

Draw *e'm* making *e'me* = B, and revolve *e'm* about *e'e*; it will generate a right cone whose rectilinear elements all make, with H, an angle equal to B. Through *s* pass the plane *sfe'* tangent to this cone, Art. (130). It, with the faces *fsd* and *sde'*, will form the required angle.

fse'' is the true size of the face opposite A, *e''* being the revolved position of *e'*, determined as in Art. (17), and the third diedral angle, formed by *e'fs* and *e'ds* may be found as in Art (52).

183. PROBLEM 52. *Given two faces of a triedral angle and their included diedral angle, to construct the angle.*

Let *dse'''* and *dsf*, Fig. 79, be the two given faces, and A the given angle.

Make *e'df* equal A. *de'* will be the vertical trace of the plane of the face *dse'''*, in its true position. Revolve *dse'''* about *sd* until *e'''* comes into *de'* at *e'*, the point where the edge *se'''*, in true position, pierces V. Draw *e'f*. It is the vertical trace of the plane of the third or required face, and *fse''* is its true size.

eon'', Art. (53), is the diedral angle opposite *e'ds*, and the third diedral angle can be found as in Art. (52).

184. PROBLEM 53. *Given one face and the two adjacent diedral angles of a triedral angle, to construct the angle.*

Let *dsf*, Fig. 79, be the given face and A and B the two adjacent diedral angles.

Make *e'df* = A. Construct as in Art. (54), *fe'*, the vertical trace of a plane *sfe'*, making with H an angle = B. SE will be the third edge. *dse'''* and *fse''* the true size of the other faces and the third diedral angle is found as in Art. (52).

185. PROBLEM 54. *Given the three faces of a triedral angle, to construct the angle.*

Let *dse'''*, Fig. 80, *dsf* and *fse''*, be the three given faces, *sd* and *sf* being the two edges in the horizontal plane.

Make *se''* = *se'''*. Revolve the face *fse''* about *fs*; *e''* describes an arc whose plane is perpendicular to H, Art. (17), of which *e''e* is the horizontal and *ee'* the vertical trace. Also revolve *dse'''* about *ds*; *e'''* describes an arc in the vertical plane. These two arcs intersect at *e'*, the point where the third edge pierces V and SE is this edge.

Join *e'd* and *e'f*. These are the vertical traces of the planes of the faces *dse'''* and *fse''*, in true position. The diedral angles may now be found as in the preceding articles.

186. PROBLEM 55. *Given the three diedral angles formed by the faces of a triedral angle, to construct the angle.*

Let A, B, and C, Fig. 81, be the diedral angles.

Make *e'df* = A. Draw *ds* perpendicular to AB and take *e'ds* as the plane of one of the faces. If we now construct a plane which shall make, with H and *e'ds*, angles respectively equal to B and C, it, with these planes, will form the required triedral angle.

To do this; with *d* as a centre and any radius as *dm*, describe a sphere; *mn'q* will be its vertical projection. Tangent to *mn'q* draw *o'u* making *o'ud* = B, and revolve it about *o'd*. It will generate a cone whose vertex is o', tangent to the sphere and all of whose rectilinear elements make with H an angle equal to B.

Also tangent to $mn'q$, draw $p'r'$ making with de' an angle equal to C, and revolve it about $p'd$. It will generate a cone whose vertex is p', tangent to the sphere, and all of whose rectilinear elements make with the plane sde' an angle equal to C. If now through o' and p' a plane be passed tangent to the sphere, it will be tangent to both cones, and be the plane of the required third face. $p'o'$ is the vertical trace of this plane, and fs tangent to the base uxy is the horizontal trace, SE is the third edge and dsf, dse''' and fse'' the three faces in true size.

187. By a reference to Spherical Trigonometry, it will be seen that the preceding six problems are simple constructions of the required parts of a spherical triangle, when either three are given. Thus in problem 50, *two sides a and c, and an angle A opposite one*, are given, and the others constructed. In problem 51, *two angles A and B, and a side b, opposite one*, are given, &c.

188. PROBLEM 56. *To construct a triangular pyramid, having given the base and the three lateral edges.*

Let cde, Fig. 82, be the base in the horizontal plane, AB being taken perpendicular to cd; and let cS, dS, and eS be the three edges.

With c as a centre and cS as a radius, describe a sphere, intersecting H in the circle mon. The required vertex must be in the surface of this sphere. With d as a centre and dS as a radius, describe a second sphere intersecting H in the circle qmp. The required vertex must also be on this surface. These two spheres intersect in a circle of which mn is the horizontal, and $m's'n'$ the vertical projection, Art. (97). With e as a centre, and eS as a radius, describe a third sphere, intersecting the second in a circle of which qp is the horizontal projection. These two circles intersect in a right line perpendicular to H at s, and vertically projected in $r's'$. This line intersects the first circle in S, which must be a point common to the three spheres, and therefore the vertex of the required pyramid. Join S with c, d and e, and we have the lateral edges, in true position.

189. PROBLEM 57. *To circumscribe a sphere about a triangular pyramid.*

Let mno, Fig. 83, be the base of the pyramid in the horizontal plane, and S its vertex.

Analysis. Since each edge of the pyramid must be a chord of the required sphere, if either edge be bisected by a plane perpendicular to it, this plane will contain the centre of the sphere. Hence, if three such planes be constructed intersecting in a point, this must be the required centre, and the radius will be the right line joining the centre with the vertex of either triedral angle.

Construction. Bisect mn and no, by the perpendiculars rc and pc. These will be the horizontal traces of two bisecting and perpendicular planes. They intersect in a right line perpendicular to H at c. Through U the middle point of SO pass a plane perpendicular to it, Art. (46). $t\mathrm{T}t'$ is this plane. It is pierced by the perpendicular (c, dc') at C, Art. (42), which is the intersection of the three bisecting planes, and therefore the centre of the required sphere. CO is its radius, the true length of which is $c'o''$, Art. (29). With c and c' as centres, and with $c'o''$ as a radius, describe circles. They will be, respectively, the horizontal and vertical projections of the sphere.

190. PROBLEM 58. *To inscribe a sphere in a given triangular pyramid.*

Let the pyramid S-mno, Fig. 84, be given as in the preceding problem.

Analysis. The centre of the required sphere must be equally distant from the four faces of the pyramid, and therefore must be in a plane bisecting the diedral angle formed by either two of its faces. Hence, if we bisect three of the diedral angles, by planes intersecting in a point, this point must be the centre of the required sphere, and the radius will be the distance from the centre to either face.

Construction. Find the angle *sps''* made by the face S*on* with H, Art. (53). Bisect this by the line *pu*, and revolve the plane *sps''* to its true position. The line *pu*, in its true position, and *on* will determine a plane bisecting the diedral angle *sps''*. In the same way determine the planes bisecting the diedral angles *srs'''*, *sqs''''*. These planes, with the base *mno*, form a second pyramid, the vertex of which is the intersection of the three planes, and therefore the required centre.

Intersect this pyramid by a plane parallel to H, whose vertical trace is *t'y'*. This plane intersects the faces in lines parallel to *mn*, *no*, and *om* respectively. These lines form a triangle whose vertices are in the edges. To determine these lines, lay off *pv* = *y'y''*, draw *vz* parallel to *ps*. *z* will be the revolved position of the point in which the parallel plane intersects *pu* in its true position. *z''* is the horizontal projection of this point, and *z''y* of the line parallel to *no*. In the same way *ty* and *tx* are determined. Draw *ox* and *ny*; they will be the horizontal projections of two of the edges. These intersect in *c* the horizontal projection of the vertex. *n'y'* is the vertical projection of the edge which pierces H at *n*, *c'* the vertical projection of the centre, and *c'd'* the radius.

With *c* and *c'* as centres describe circles with *c'd'* as a radius. They will be the horizontal and vertical projections of the required sphere.

PART II.

SPHERICAL PROJECTIONS.

PRELIMINARY DEFINITIONS.

191. One of the most interesting applications of the principles of Descriptive Geometry is to *the representation, upon a single plane, of the different circles of the earth's surface*, regarded as a perfect sphere.

These representations are *Spherical Projections.* The plane of projection which is generally taken as that of one of the great circles of the sphere, is *the primitive plane;* and this great circle is *the primitive circle.*

The axis of the earth is the right line about which the earth is known daily to revolve.

The two points in which it pierces the surface are *the poles*, one being taken as *the North* and the other as *the South Pole.*

The axis of a circle of the sphere is the right line through its centre perpendicular to its plane. and the points in which it pierces the surface are *the poles of the circle.*

The polar distance of a point of the sphere is its distance from either pole of the primitive circle.

The polar distance of a circle of the sphere is the distance of any point of its circumference from either of its poles.

192. The lines on the earth's surface, usually represented, are:

1. *The Equator*, the circumference of a great circle whose plane is perpendicular to the axis.

2. *The Ecliptic*, the circumference of a great circle making an angle of 23½°, nearly, with the equator. It intersects the equator in two points, called *the Equinoctial Points.*

3. *The Meridians*, the circumferences of great circles whose planes pass through the axis; and are therefore perpendicular to the plane of the equator.

Of these meridians two are distinguished: the *Equinoctial Colure*, which passes through the equinoctial points; and the *Solstitial Colure*, whose plane is perpendicular to that of the equinoctial colure.

The solstitial colure intersects the ecliptic in two points, called *the Solstitial Points.*

4. *The Parallels of Latitude*, the circumferences of small circles parallel to the equator.

Four of these are distinguished:

The Arctic Circle, 23½° from the north pole;
The Antarctic Circle, 23½° from the south pole
The Tropic of Cancer, 23½° north of the equator;
The Tropic of Capricorn, 23½° south of the equator.

The first two are also called *Polar Circles.*

193. *The Latitude* of a point on the earth's surface, is its distance from the equator, measured on a meridian passing through the point.

The Horizon of a point or place, on the earth's surface, is the circumference of a great circle whose plane is perpendicular to the radius passing through the point.

Let M, Fig. 85, be any point on the earth's surface. Through this point and the axis pass a plane, and let MNS be the circumference cut from the sphere; N the north and S the south pole ECE′ the intersection of the plane with the equator, and PCQ perpendicular to CM, its intersection with the horizon of the given point. Then ME is the latitude of the point, and NQ the distance of the pole N from the horizon. NQ also measures the angle NCQ, the inclination of the axis to the horizon. But

$$NQ = ME,$$

since each is obtained by subtracting NM from a quadrant; that is, *the distance from either pole of the earth to the horizon of a place, is equal to the latitude of that place.*

194. Let NS, Fig. 85, and MR be the axes of two circles intersecting the plane NCM in the lines EE′ and PQ respectively. ECP is then the angle made by the planes of these circles. But

$$ECP = NCM,$$

since each is obtained by subtracting MCE from a right angle; that is, *the angle between any two circles of the sphere, is equal to the angle formed by their axes.*

195. If a plane be passed through the axes of any circle of the sphere and of the primitive circle, its intersection with the primitive plane is *the line of measures of the given circle.* This auxiliary plane is perpendicular to the planes of both circles, and therefore to their intersection; hence the line of measures, a line of this plane, must be *perpendicular to the intersection* of the given with the primitive circle, and must also *pass through the centre* of the primitive circle. Thus, if EE′, Fig. 85, is the intersection of a circle with the primitive plane NESE′, NS is its line of measures. Also NS is the line of measures of any small circle whose inter section with the primitive plane is parallel to EE′.

ORTHOGRAPHIC PROJECTIONS OF THE SPHERE.

196. When the point of sight is taken in the axis of the primitive circle, and at an infinite distance from this circle, *the projections of the sphere are orthographic*, Art. (2).

If E, Fig. 85, be any point, e will be its orthographic projection on the plane of a circle whose axis is CM. But

$$Ce = Ed;$$

that is, the orthographic projection of any point of the surface of a sphere is *at a distance from the centre of the primitive circle equal to the sine of its polar distance.*

197. The circumference of a circle, oblique to the primitive plane, *is projected into an ellipse.* For the projecting lines of its different points form the surface of a cylinder whose intersection with the primitive plane is its projection, Art. (74), and this intersection is an ellipse, Art. (160).

If the plane of the circumference be perpendicular to the primitive plane, its projection is a right line, Art. (62).

If the plane of the circumference be parallel to the primitive plane, its projection is an equal circumference, Art. (62).

The projection of every diameter of the circle which is oblique to the primitive plane, will be a right line less than this diameter, Art. (29), while the projection of that one which is parallel to the primitive plane, will be equal to itself, Art. (14). This projection will then be longer than any other right line which can be drawn in the ellipse, and is therefore its *transverse axis*. Analyt. Geo., Art. (127).

The projection of that diameter which is perpendicular to the one which is parallel to the primitive plane, will be perpendicular to this transverse axis, Art. (36), and pass through the centre, and therefore be *the conjugate axis* of the ellipse, Art. (59).

This last diameter is perpendicular to the intersection of the plane of the given circle and the primitive plane, and therefore makes with the primitive plane the same angle as the circle; and one-half its projection, or *the semi-conjugate axis* of the ellipse, is evidently *the cosine of this inclination*, computed to the radius of the given circle. Hence, to project any circle orthographically, we have simply *to find the projection of that diameter which is parallel to the primitive plane, and through its middle point draw a right line perpendicular to it, and make it equal to the cosine of the angle made by the circle with the primitive plane.* The first line is the transverse, and the second the semi-conjugate axis of the required ellipse, which may then be accurately constructed as in Art. (59).

It should be remarked that the conjugate axis of the ellipse always lies on the line of measures of the circle to be projected, Art. (195).

198. The line of measures of a circle evidently contains the projections of both poles of the circle, Art. (195) ; and since the arc which measures the distance of either pole from the pole of the primitive circle, measures also the inclination of the two circles, Art. (194), it follows that either pole of a circle is orthographically projected in its line of measures, *at a distance from the centre of the primitive circle equal to the sine of its inclination*, Art. (196).

199. Problem 59. *To project the sphere upon the plane of any one of its great circles.*

Let EpE$'q$, Fig. 86, be the primitive circle intersecting the equator in the points E and E$'$, and making with it an angle denoted by A. Let E and E$'$ also be assumed as the equinoctial points. The line EE′ is then the intersection of the primitive plane by the equator, ecliptic, and equinoctial colure, and pq perpendicular to it is the line of measures of all these circles.

Let us first project the hemisphere lying between the primitive plane and the north pole.

Since EE′ is that diameter of *the equator* which lies in the primitive plane, it is its own projection, and therefore the transverse axis of the ellipse into which the equator is projected. From q lay off $qm' = \text{A}$, and draw $m'm$ perpendicular to pq. $\text{C}m = \cos \text{A}$, and is the semi-conjugate axis, Art. (197). On this and EE′ describe the semi-ellipse EmE′; it is the projection of that part of the equator lying above the primitive plane.

n is the projection of the *north pole*, En' being made equal to A, and $\text{C}n = n'x$ its sine, Art. (198).

EE′ is also the transverse axis of the projection of *the ecliptic.* If the portion of the ecliptic on the hemisphere under consideration, lies between the equator and the north pole, it will make an angle with the primitive plane greater than that of the equator by $23\frac{1}{2}°$. If it lies between the equator and the south pole, it will make a less angle by $23\frac{1}{2}°$. Taking the former case, lay off $m'o' = 23\frac{1}{2}°$, then $qo = \text{A} + 23\frac{1}{2}°$, and $\text{C}o = \cos(\text{A} + 23\frac{1}{2}°) =$ the semi-conjugate axis, with which and EE′ describe the semi-ellipse EoE′, the projection of one half the ecliptic.

The equinoctial colure, making with the equator a right angle, makes with the primitive plane an angle equal to $90° + \text{A}$. EE′ is the transverse axis of its projection, and $\text{C}n = \cos qn' = \cos(90° + \text{A}) =$ the semi-conjugate axis. And the semi-ellipse EnE′ is the projection of that half above the primitive plane.

The solstitial colure being perpendicular to the equator and equinoctial colure is perpendicular to EE′, and therefore to the primitive plane; hence its projection is the right line qp, Art. (197).

To project *any meridian*, as that which makes with the solstitial colure, an angle denoted by B; pass a plane tangent to the sphere at the north pole. It will intersect the planes of the given meridian and solstitial colure in lines perpendicular to the axis and making with each other an angle equal to B; and these lines will pierce the primitive plane in points of the intersections of the planes of these meridians with the primitive plane, Art. (30). To determine this tangent plane; revolve the solstitial colure about pq, as an axis, until it comes into the primitive plane. It will then coincide with E$n'pq$, and the north pole will fall at n'. Draw $n't$ tangent to E$n'p$; it is the revolved position of the intersection of the required tangent plane by the plane of the solstitial colure. It pierces the primitive plane at t, and st perpendicular to Cn, Art. (145), is the trace of the tangent plane. Revolve this plane about ts until it coincides with the primitive plane. The north pole falls at n''; tn'' being equal to tn', Art. (17). Through n'', draw $n''s$, making with $n''t$ an angle equal to B. This will be the revolved position of the intersection of the tangent plane by the plane of the given meridian. It pierces the primitive plane at s, and sC is the intersection of the meridian plane with the primitive, and yz is the transverse axis of the required projection, Art. (197).

To find the semi-conjugate; through the north pole pass a plane perpendicular to yz; nv is its trace. Revolve this plane about nv until it coincides with the primitive plane. N falls at n''', and $n'''vn$ is the angle made by the meridian with the primitive plane, Art. (53). Lay off $vk =$ CE, and draw ku parallel to yz. Cu is the required semi-conjugate axis, and the semi-ellipse yuz is the projection of that half of the meridian which lies above the primitive plane.

Since the plane of *any parallel of latitude*, as the arctic circle, is parallel to the equator, it will be intersected by the plane of the equinoctial colure in a diameter parallel to EE′, and to the primitive plane, and the projection of this diameter will be the transverse axis of the projection. To determine it; revolve the equinoctial colure about EE′ as an axis until it coincides with EpE′q; N falls at p. From p lay off $pa' = 23\frac{1}{2}°$, the polar distance of the parallel, and draw $a'b'$. It will be the revolved position of that diameter of the arctic circle which is parallel to the primitive plane. When the colure is revolved to its true position $a'b'$ will be projected into ab, the required transverse axis. From d its middle point lay off $di = \cos A$, computed to the radius da; it will be the semi-conjugate axis, and the ellipse $aibh$ is the required projection.

In the same way the tropic of Cancer or any other parallel may be projected.

If the polar distance of the parallel is greater than $90° - A$, the inclination of the axis, the parallel will pass below the primitive plane and a part of its projection be drawn broken.

The projection of the tropic of Cancer is tangent to EoE′ at o, Art. (65).

200. Each point in the primitive circle is evidently the projection of two points of the surface of the sphere, one above and the other below the primitive plane. To represent these points distinctly and prevent the confusion of the drawing, we first project the upper hemisphere, as above, and then revolve the lower 180°, about a tangent to the primitive circle at E. It will then be above the primitive plane and may be projected in the same way as the first. Em''E″ is the projection of the other half of the equator; s of the south pole; Eo''E″ of the other half of the ecliptic; EsE″ of the equinoctial colure; $y'sz'$ of the meridian, &c.

201. *If the projection be made on the equator*, the preceding problem is much simplified. Thus, let EpE′q, Fig. 87, be the equator. n is the projection of the north pole; EoE′ of one half of the ecliptic, qo' being equal to $23\frac{1}{2}°$, and no its cosine.

Since the meridians are all perpendicular to the equator, EE′ is the projection of the equinoctial, and pq of the solstitial colure; yz of the meridian making an angle of 30° with the solstitial colure.

Since the parallels of latitude are parallel to the primitive plane, *abbi* is the projection of the arctic circle, and *ogkl* that of the tropic of Cancer, *na* being equal to the sine of $23\frac{1}{2}°$, and $nl =$ sin $66\frac{1}{2}°$, Art. (196). *ogkl* is tangent to E*o*E′ at *o*.

202. *If the projection be made on the equinoctial colure;* let ENE′S, Fig. 88, be the primitive circle, and E and E′ the equinoctial points.

Since the equator is perpendicular to the primitive plane, EE′ will be its projection. N is the north and S the south pole. E*o*E′ is the projection of one half the ecliptic; C*o* being equal to cos $66\frac{1}{2}°$. NS is the projection of the solstitial colure.

Since the parallels are perpendicular to the primitive plane, *ab* and *a′b′* are the projections of the polar circles; N*a* and S*a′* being each equal to $23\frac{1}{2}°$; and *lg* and *l′g′* the projections of the tropics. N*g* and S*g′* being each equal to $66\frac{1}{2}°$.

N*y*S is the projection of one half the meridian, making an angle of 30° with the solstitial colure, or 60° with the primitive plane. C*y* being equal to $\cos 60° = \frac{1}{2}$ CE′.

203. The projections may be made upon the ecliptic, and horizon of a place, in the same way as in problem 59. In the former case, the angle A will equal $23\frac{1}{2}°$; and in the latter, since the angle included between the axis and horizon is equal to the latitude of the place, Art. (193), the angle A between the equator and horizon will be 90° + the latitude.

STEREOGRAPHIC PROJECTIONS OF THE SPHERE.

204. The natural appearance and beauty of a scenographic drawing will depend, very much, upon the position chosen by the draughtsman, or artist, for the point of sight. This should be so selected that a person taking the drawing into his hand for exam

ination, will naturally place his eye at this point. From any other position of the eye, the drawing will appear to some extent distorted. Hence it is, that an orthographic drawing never appears perfectly natural, as it is impossible to place the eye of the observer at an infinite distance from it.

In spherical projections, *if the point of sight be taken at either pole of the primitive circle*, the projections are *Stereographic*, and, in general, present the best appearance to the eye of an ordinary observer, as, in this case, the projections of all circles of the sphere, as will be seen in Art. (207), are circles.

205. The projection of each point on the surface of the sphere, will be that point in which a right line, through it and the point of sight, pierces the primitive plane, Art. (3).

Let M, Fig. 89, be any point on the surface of the sphere. Through it and the axis of the primitive circle pass a plane. It will intersect the sphere in a great circle EME'S, and the primitive plane in a right line EE'; N and S being the poles of the primitive circle and S the point of sight. NM is the polar distance and *m* the stereographic projection of M. C*m* is the tangent of the arc C*o*, computed to the radius CS = CE, and C*o* is one half of NM. That is, the stereographic projection of any point of the surface of the sphere is *at a distance from the centre of the primitive circle equal to the tangent of one half its polar distance.*

In this projection, it should be observed that the polar distance of a point is always its distance from the pole opposite the point of sight, and often exceeds 90°.

206. If a plane be passed through the vertex of an oblique cone with a circular base, perpendicular to this base and through its centre, such plane is *a principal plane*, and evidently bisects all chords of the cone drawn perpendicular to it.

Let SAB, Fig. 90, be such a plane intersecting the cone in the elements SA and SB, and the base in the diameter AB. If this cone be now intersected by a plane tTt', perpendicular to the principal plane, and making with one of the principal elements, as SA, an angle Sba equal to the angle SBA, which the other makes with the plane of the base, the section is a *sub contrary section* and will be the circumference of a circle. For, through o, any point of ba, which is the orthographic projection of the curve of intersection on the principal plane, pass a plane parallel to the base. It cuts from the cone the circumference of a circle, and intersects the plane of the sub contrary section in a right line perpendicular to SAB at o, and the two curves have, at this point, a common ordinate. The similar triangles aod and cob give the proportion

$$ao : oc :: od : ob; \quad \text{or,} \quad ao \times ob = oc \times od.$$

But $oc \times od$ is equal to the square of the common ordinate, since the parallel curve is a circle; hence $ao \times ob$ is equal to the square of the ordinate of the sub contrary section, which must, therefore, be *a circle.*

207. *To project any circle of the sphere;* through its axis and the axis of the primitive circle pass a plane, and let ENE'S, Fig. 89, be the circle cut from the sphere by this plane; S the point of sight; RM the orthographic projection of the given circle on the cutting plane, Art. (62); CN the axis of the primitive circle orthographically projected in EE', and CP' the axis of the given circle.

The projecting lines, drawn from points of the circumference to S, form a cone whose intersection by the primitive plane will evidently be the stereographic projection of the circumference, SRM is the principal plane of this cone, and SR and SM the principal elements. The primitive plane is perpendicular to this plane, and intersects the cone in a curve of which rm is the orthographic projection. But the angle

$$\text{S}rm = \text{SMR}.$$

Since each is measured by SR, ES being equal to E'S, hence this section is a sub contrary section, and therefore a circle whose diameter is mr. That is, *the stereographic projection of every circle on the surface of a sphere*, whose plane does not pass through the point of sight, *is a circle.*

mr is also the line of measures of the given circle, Art. (195) and evidently contains *the centre of its projection.*

The distance

$$\text{C}r = \tan \text{C}o' = \tan \tfrac{1}{2}\,(\text{PR} + \text{PN}),$$

and

$$\text{C}m = \tan \text{C}o = \tan \tfrac{1}{2}\,(\text{PR} - \text{PN}).$$

Hence, the extremities of *a diameter of the projection* of any circle, on the surface of the sphere, are in its line of measures, one *at a distance from the centre of the primitive circle, equal to the tangent of one-half the sum of the polar distance and inclination* of the circle, and the other at a distance equal to *the tangent of one-half the difference of these two arcs.*

When the polar distance is greater than the inclination, these extremities will evidently be on different sides of the centre of the primitive circle. When less, they will be on the same side. If the polar distance is equal to the inclination, the projection of the given circle will pass through the centre of the primitive circle.

The polar distance and inclination of any circle being known, a diameter of its projection can thus be constructed, and thence the projection.

208. If the circle be parallel to the primitive plane, the sub

contrary and parallel sections coincide, and the projection is a circle whose centre is at the centre of the primitive circle, and radius the distance of the projection of any point of the circumference from the centre of the primitive circle; that is, *the tangent of half the circle's polar distance*, Art. (205).

If the plane of the circle pass through the point of sight, the projecting cone becomes a plane, and *the projection is a right line.*

209. *If a right line be tangent to a circle of the sphere, its projection will be tangent to the projection of the circle.* For, the projecting lines of the circumference form a cone, and those of the tangent, a plane tangent to this cone, along the projecting line of the point of contact; hence the intersections of the cone and plane by the primitive plane, are tangent to each other at the projection of the point of contact, Art. (112). But the first is the projection of the circle, and the second that of the tangent.

210. Let MR and MT, Fig. 91, be the tangents to two circles of the sphere at a common point, M. Let these tangents be projected on the primitive plane by the planes RMS and TMS respectively, in the lines *mr* and *mt*, and let M*a*S and M*b*S be the circles cut from the sphere by these planes, and let SR and ST be the lines cut from the tangent plane to the sphere at S. Since this tangent plane is parallel to the primitive plane, the lines SR and ST will be parallel respectively to *mr* and *mt*, and the angle RST $=$ *rmt*. Join RT. Since RS and RM are each tangent to the circle M*a*S, they are equal, and for the same reason TM $=$ TS; hence the two triangles, RMT and RST, are equal, and the angle

$$\text{RMT} = \text{RST} = rmt,$$

that is, *the angle between any two tangents to circles of the sphere, at a common point, is equal to the angle of their projections.*

The angle between the circles is the same as that between their tangents, and since the projections of the tangents are tangent to the projections of the circles, the angle between the projections of the circles is the same as that between the projections of the tan gents; hence *the angle between any two circumferences or arcs is equal to the angle between their projections.*

211. *If from the centres of the projections of two circles radii be drawn to the intersection of these projections, they will make the same angle as the circles in space.* For, these radii being perpendicular to the tangents to the projections, at their common point, make the same angle as these tangents, and therefore as the projections of the arcs, or as the arcs themselves.

212. If the circle to be projected be *a great circle*, it will intersect the primitive circle in a diameter perpendicular to its line of measures, Art. (195). Let O, Fig. 92, be the centre *of the projection* of such a circle intersecting the primitive circle EPE'R in the diameter PR, CE being its line of measures, and P and R evidently points of the projection. Draw the radius OR. The primitive circle is its own projection; therefore the angle CRO is equal to the angle between the given and primitive circles, Art. (211). CO is the tangent of this angle, and OR its secant. Hence *the centre of the projection of a great circle*, is in its line of measures, Art. (207), at a distance from the centre of the primitive circle equal to *the tangent of its inclination*, and the radius of the projection is *the secant of this angle.*

213. Let O, Fig. 93, be the centre *of the projection of a small circle* perpendicular to the primitive plane, and intersecting it in PR. OC is its line of measures, and P and R points of the projection. Join CR and OR. OR is perpendicular to CR, since EPE'R is the projection of the primitive circle, and P*p*R that of

the given circle, at right angles with it, Art. (211). OR is therefore the tangent of the arc ER, the polar distance of the given circle, and CO is its secant. Hence *the centre of the projection of a small circle, perpendicular to the primitive plane*, is in its line of measures, at a distance from the centre of the primitive circle equal to *the secant of the polar distance*, and the radius of the projection is *the tangent of the polar distance.*

214. Let P and R, Fig. 94, be the poles of any circle of the sphere; EPE'S being the circle cut from the sphere by the plane of the axes of the given and primitive circles, and MQ, the intersection of this plane with that of the given circle, and EE', its intersection with the primitive circle, the line of measures of the given circle. p is the projection of P and r of R. Cp is the tangent of Co, equal to one-half of NP, which measures the inclination of the given circle to the primitive, Art. (194). Cr is the tangent of Co', the complement of Co, or is the co-tangent of Co. Hence *the poles of any circle* of the sphere are projected into the line of measures, the one furthest from the point of sight at a distance from the centre of the primitive circle equal *to the tangent of half the inclination*, and the other at a distance equal to *the co-tangent of half the inclination* of the given to the primitive circle.

215. Problem 60. *To project the sphere upon the plane of any of its great circles, as the ecliptic.*

Let EpE'q, Fig. 95, be the primitive circle intersecting the equator in EE'. In this case, E and E' will be the equinoctial points, and in any other case may be taken as such. EE' will also be the intersection of the plane of the equinoctial colure with the primitive plane, and pq the line of measures of both these circles.

We will first project the hemisphere above the primitive plane, the point of sight being at the pole underneath.

Since *the Equator* makes an angle of $23\frac{1}{2}°$ with the primitive plane, we draw E$'o$, making the angle CE$'c = 23\frac{1}{2}°$. Co is the tangent of this angle and E$'o$ the secant; hence with o as a centre and E$'o$ as a radius, describe the arc EmE$'$; it is the projection of the part of the equator above the primitive plane, Art. (212).

From E lay off Ek $= 23\frac{1}{2}°$ and draw kE$'$; Cn is the tangent of half the inclination of the equator, and n is the projection of *the north pole*, Art. (214).

The *Equinoctial colure* makes with the primitive plane an angle $= 90° + 23\frac{1}{2}°$. Through E$'$ draw E$'x$ perpendicular to E$'o$. The angle CE$'x = 90° + 23\frac{1}{2}°$, and C$r$ = tan CE$'r$ is its tangent and E$'r$ its secant. With r as a centre, and E$'r$ as a radius describe E$'n$E. It is the projection of the half of the equinoctial colure above the primitive plane. It must pass through n.

The Solstitial Colure, being perpendicular to EE$'$, passes through the point of sight and is projected into the right line pq.

To project any other meridian, as that which makes an angle of 30° with the equinoctial colure; produce the arc E$'n$E until it intersects Cr produced. The point of intersection, which we denote by s, will be the projection of the south pole, and since all the meridians pass through the poles, their projections will pass through n and s, and ns will be a chord common to the projections of all the meridians. If at its middle point r, the perpendicular rl be erected, this will contain the centres of all these projections. If through n, nl be drawn making $rnl = 30°$, it will be the radius of the projection of that meridian which makes an angle of 30° with the equinoctial colure, since rn is the radius of the projection of this colure Art. (211); and l is the centre of the projection of the required meridian, and ynz the projection.

To project a parallel of latitude, as *the Arctic Circle;* lay off E$i = 47°$, the sum of the inclination, $23\frac{1}{2}°$, and the polar distance $23\frac{1}{2}°$ Art. (207). Draw E$'i$; Co = tan $\frac{1}{2}$Ei, and o is one extremity of a diameter of the projection. Since the inclination is equal to the polar distance, the other extremity is at C, Art. (207) and the circle on Co, as a diameter, is the required projection.

For the Tropic of Cancer; lay off $Ep = 23\frac{1}{2}° + 66\frac{1}{2}°$; Cp is the tangent of half Ep, and p one extremity of a diameter of the projection. From k lay off $kh = 66\frac{1}{2}°$ the polar distance of the parallel. Then $Eh = 43°$ equal the difference between the polar distance and the inclination, and Cv is the tangent of its half, and v the other extremity of the diameter, and the circle on pv, the required projection. The projection of this tropic is tangent to the ecliptic at p, Art. (65).

216. Since each point on the hemisphere, below the primitive plane, has a greater polar distance than 90° and will therefore be projected without the primitive circle, Art. (205), and those circles near the point of sight will thus be projected into very large circles, we make a more natural representation of this hemisphere by revolving it 180°, as in the orthographic projection, about a tangent at E, the point of sight being moved to the pole of the primitive circle in its new position. The hemisphere is then above the primitive plane and is projected as in the preceding article; s being the projection of the south pole; $Em''E''$ of the other half of the equator; EsE'' of the equinoctial colure, &c.

217. If the projection be made on any other great circle than the ecliptic, as on that making with the equator an angle denoted by A, the construction will be the same, the angle A being used instead of $23\frac{1}{2}°$.

If on the horizon of a place, A must equal 90° minus the latitude. Art. (193).

218. *If the projection be made on the equator*, the preceding problem is much simplified. Thus let $EpE'q$, Fig. 96, be the equator, E and E′ the equinoctial points. EE′ is the intersection of the plane of the ecliptic with that of the equator and pq is its

line of measures and EoE′ its projection, m being the centre and $mn = \tan 23\frac{1}{2}°$.

Since *the meridians* pass through the point of sight they are projected into right lines. EE′ is the projection of the equinoctial and pq of the solstitial colure; and ynz of the meridian which makes an angle of 30° with the solstitial colure, n being the projection of the north pole.

The parallels of latitude, being parallel to the equator, are projected as in Art. (208); *the Arctic Circle* into arb; and *the Tropic of Cancer* into dof; na being equal to $\tan \frac{1}{2}(23\frac{1}{2}°)$; and $nd = \tan \frac{1}{2}(66\frac{1}{2}°)$.

219. *If the projection be on the solstitial colure;* let ENE′S, Fig. 97, be the primitive circle; EE′ its intersection with the plane of the equator.

The Equator, being perpendicular to the primitive plane, passes through the point of sight and is projected into EE′.

The Ecliptic, for the same reason, is projected into oo', oCE being equal to $23\frac{1}{2}°$.

NyS is the projection of *the meridian*, making with the primitive plane an angle of 30°, m being its centre and Cm = tan 30°.

adb and $a'hb'$ are the projections of the *polar circles*, Cx and Cx' being each equal to the secant of $23\frac{1}{2}°$, and xb and $x'b'$ each equal to the tangent of $23\frac{1}{2}°$, Art. (213).

ogf and $o'g'd'$ are the projections of *the tropics*, each described with a radius equal to the tan $66\frac{1}{2}°$.

GLOBULAR PROJECTIONS.

220. By an examination of an orthographic or stereographic projection, it will be observed that the projections of equal arcs, of great circles which pass through the pole of the primitive circle, are very unequal in length. In the orthographic, as the arc is re-

moved from the pole its projection is diminished, and when near the primitive circle becomes very small, while the reverse is the case in the stereographic.

To avoid this inequality, as far as possible, the point of sight is taken *in the axis of the primitive circle, without the surface, and at a distance from it equal to the sine of* $45° = R\sqrt{\frac{1}{2}}$.

Spherical projections, with this position of the point of sight are called *Globular*.

Thus let the quadrant Ep, Fig. 98, be bisected at M, and S the point of sight. M is projected at m, and Cm will be equal to mE. For we have the proportion :

$$oS : oM :: CS : Cm; \quad \text{whence,} \quad Cm = \frac{CS \times oM}{oS} \cdots (1).$$

Since

$$oM = oC = qS = R\sqrt{\tfrac{1}{2}},$$

we have

$$oS = R + 2R\sqrt{\tfrac{1}{2}}, \qquad \text{and} \qquad CS = R + R\sqrt{\tfrac{1}{2}}$$

Substituting these in (1), and reducing, we have

$$Cm = \frac{R}{2} = mE;$$

and it will be found that the projections of any other two equal arcs of this quadrant are very nearly equal. This is the only advantage of this mode of projection, as the projections of the circles of the sphere, being the intersections of their projecting cones with the primitive plane will, in general, be ellipses.

GNOMONIC PROJECTION.

221. If the sphere be projected on a tangent plane at any point, *the point of sight being at its centre*, the projection is called *Gnomonic.*

In this case the projections of all meridians are right lines, since their planes pass through the point of sight.

If the point of contact be on the equator the projections of the parallels of latitude will be arcs of hyperbolas, Art. (165).

If the point of contact be at either pole of the earth, these projections will be circles.

By this mode of projection, the portions of the sphere distant from the point of contact will be very much exaggerated.

CYLINDRICAL PROJECTION.

222. If a cylinder be passed tangent to a sphere along the equator, and the point of sight be taken at the centre of the sphere, and the circles of the sphere be projected on the cylinder, and the cylinder be then developed, we have a developed projection called *the cylindrical projection.*

In this case the meridians will be projected into right lines, elements of the cylinder, which are developed into parallel lines perpendicular to the developed equator, Art. (161); and the parallels into circles which are developed into right lines perpendicular to the developed meridians, and at distances from the equator each equal to the tangent of the latitude of the parallel.

CONIC PROJECTION.

223. If a cone be passed tangent to a sphere along one of its parallels of latitude, and the circles of the sphere be projected on it, the point of sight being at the centre, and the cone be then developed, we have a developed projection called *the conic projection.*

In this case the meridians will be projected into right lines.

elements of the cone, which are developed into right lines passing through the vertex; and the parallels into circles whose developments will be arcs described from the vertex, each with a radius equal to the distance of the projection from the vertex, Art. (166.)

Thus, in Fig. 99, let EPE′ be a section of the sphere and V the vertex of the cone tangent along the parallel of which *ab* is the orthographic projection. The equator will be projected into a circle whose diameter is *ee′* and the parallel MN into one whose diameter is *mn*, the first being developed into an arc of which V*e* is the radius, and the second into one of which V*m* is the radius. The radius V*a* of the development of the circle of contact is evidently the tangent of its polar distance.

The drawing in this, as in the cylindrical projection, for those portions of the sphere distant from the circle of contact, will evidently greatly exaggerate the parts projected.

224. This exaggeration may be lessened by making the projection on a cone passing through two circumferences equally distant, one from the equator and the other from the pole. Thus, let *ab* and *cd*, Fig. 100, represent two circles equally distant from EE′ and P, E*a* being one fourth of the quadrant; in this case while the parts E*a* and *c*P will be exaggerated in projection, the part *ac* will be lessened.

225. When a small portion of the surface, between two given parallels, is to be represented, the conic method may be well used, taking the cone tangent to the sphere along a parallel midway between the given parallels, if the first method be used; or passing through two parallels, each distant from a limiting parallel one fourth the arc of the given portion, if the second method b used.

CONSTRUCTION OF MAPS.

226. If it be desired to represent the entire surface of the earth by a map, either of the preceding methods may be used. In this case it is usual to divide the quadrant, from the pole to the equator, into *nine* equal parts and project the parallels of latitude through each of these points, as well as the polar circles and tropics; and also to divide the semi-equator into *twelve* equal parts and to project the meridians passing through these points. These meridians will be 15° apart and are called *hour-circles.*

The projections of the different points to be represented are then made and the map filled up in detail.

227. The Stereographic projection gives the most natural representation and, in general, is of the easiest construction.

In the Globular, when the equator is taken as the primitive circle, the projections of the meridians are right lines, and of the parallels of latitude, circles; and this projection has the advantage that these parallels, which are equally distant in space, have their projections also very nearly equally distant.

228. A very simple construction, when the primitive circle is a meridian, is sometimes made thus: Divide the arcs EN, E′N, ES, and E′S, Fig. 101, each into *nine* equal parts, and the radii CN and CS also each into nine equal parts, then describe arcs of circles through each of the three corresponding points of division, for the representatives of the parallels.

In the same way divide CE and CE′, each into *six* equal parts, and, through the points of division and the poles N and S, describe arcs of circles for the representatives of the meridians.

This representation, though called the *equidistant projection,* is not strictly a projection. It differs little, however, from the globular projection.

LORGNA'S MAP.

229. A map of the globe is sometimes made by describing a primitive circle with a radius equal to $R\sqrt{2}$, R being the radius of the sphere, and regarding this as the representative of the equator. Through its centre draw right lines making angles with each other of 15°, for the meridians.

To represent the parallels: With the centre of the primitive circle as a centre and with a radius equal to $\sqrt{2Rh}$, h being the altitude of the zone, included between the pole and the parallel, describe a circle to represent each parallel in succession.

The area of the primitive circle is evidently equal to the area of the hemisphere, and the area of each other circle to that of the corresponding zone. Hence, the area between any two circumferences will be equal to that of the zone included between the corresponding parallels.

Also the area of any quadrilateral formed by the arcs of two meridians and two parallels will be equal to its representative on the primitive plane.

MERCATOR'S CHART.

230. This chart, which is much used by navigators, is a modification of the cylindrical projection. It has the great advantage, that the course of a ship on the surface of the sphere which makes a constant angle with the meridians intersected by it, will be represented on the chart by a right line.

To secure this, as the length of the representative of a degree of longitude, as compared with the arc itself, is manifestly increased as the distance from the equator increases, it is necessary that the representatives of the degrees of latitude should increase in the same ratio.

But the length of a degree of longitude, at any latitude, is known to be equal to the length of a degree at the equator multiplied by the cosine of the latitude, and since the representative of this degree at all latitudes is constant on the chart, being the distance between two parallel lines the representatives of two consecutive meridians, it follows that as we depart from the equator, this representative, as compared with the arc itself, increases as the cosine of the latitude decreases, or increases as the secant increases, and hence the representative of the degree of latitude must increase in the same ratio; that is, this representative, at any given latitude, must equal *its length at the equator multiplied by the natural secant of this latitude.*

By adding the representatives of the several degrees, or, still more accurately, of the several minutes of a quadrant, the distance of the representative of each parallel from the equator may be found, and the chart may then be thus constructed. Draw a right line to represent the equator, then a system of equidistant parallel lines for the meridians; on either one of these lay off the proper distances computed as above for each parallel to be represented, and through the extremities of these distances draw right lines perpendicular to the system first drawn. They represent the parallels.

231. All the maps constructed as in the preceding articles, though giving a general representation of the relative position of objects on the earth's surface, are defective in this, that there is no definite relation between actual distances of points and the representation of these distances on the maps, so that there can be no scale on the map by which these actual distances can be determined.

As in detailed representations of smaller portions of the surface, this scale is absolutely necessary, other modes of constructing these maps have been devised, by which a near approximation to an accurate scale is made.

FLAMSTEAD'S METHOD.

232. In this method, modified and improved, and now in very general use, a right line, AB, Fig. 102, is drawn, which represents the rectified arc of the meridian passing through the middle of the portion to be mapped. A point, c, is then assumed to represent the point in which the parallel midway between the extreme parallels intersects this meridian, and from this point, in both directions, equal distances, cx, cy, &c., are laid off, each representing the true length of one degree of the meridian. Then with a radius equal to the tangent of the polar distance of this central parallel, the arc dce is described to represent the parallel. This arc is the development of the line of contact of a cone tangent to the sphere. Also, with the same centre, arcs are described through each of the points of division x, y --- r, o, to represent the other parallels one degree distant from each other. Then, on each of these arcs, from AB, lay off both ways the arcs ca, ca', yv, yv', os, os', &c., each equal to the length of a degree of longitude at the points c, o, &c., viz, the length of a degree at the equator multiplied by the natural cosine of the latitude of the point. Through the points a, v, s, &c., and a', v', s', &c., draw the lines avs, and $a'v's'$. They will represent the two meridians making an angle of one degree each, with the central meridian.

In the same way, the representatives of the two meridians next to these may be constructed, by laying off on the same arcs from a, v, s, &c., distances each equal to ca, yv, os, &c., and drawing lines through the points thus determined, and so on, until the representatives of the extreme meridians of the portion to be mapped are drawn.

In this map the scale along the central meridian and parallels will be accurate. In other directions when the map does not rep resent a very extended portion of the sphere, a very near approximation to accuracy is made.

THE POLYCONIC METHOD.

233. In this method, by which the elegant and accurate maps of the U. S. Coast Survey are constructed, the central meridian and parallel are represented as in the preceding article. The representatives of the other parallels are each described, through the proper point of division, with a radius equal to the tangent of its polar distance, thus giving the development of the lines of contact of so many tangent cones to the sphere. The length of each degree of longitude is then laid off as in the former method and the representatives of the meridians drawn.

In this method, also, the scale, along the central meridian and parallels, is accurate and in other directions very nearly so. This has the advantage that the representatives of the meridians and parallels are perpendicular to each other as in space, which is not the case in Flamstead's method.

234. When very small areas are mapped this method is thus modified in the coast survey office. The same process is used as above to construct the representatives of the meridians with accuracy. Then commencing with the central parallel, the distance *cx*, Fig. 102, between it and the consecutive one, as measured on AB, is laid off on each meridian in both directions and through the extremities lines drawn to represent the consecutive parallels. Then from these the same distances are laid off for the next and so on until all are constructed.

The first set of parallels, described as in the preceding article, which should be in pencil, are then erased.

By this method equal meridian distances are everywhere included between the parallels, and the scale accurate only in the direction of the meridians, and central parallel. This is called the *equidistant polyconic method.*

235. When the polar distance of the parallel is much greater than 45°, the practical construction of its representative becomes difficult, as its centre will be so far distant. In such case, for both the polyconic method and that of Flamstead, tables are carefully computed giving the rectilineal co-ordinates of the points of the representatives of the parallels, for each minute of latitude and longitude, and these representatives can then be accurately constructed by points. The tables thus computed in the U. S. Coast Survey office are very much extended and of great value.

PART III.

SHADES AND SHADOWS.

PRELIMINARY DEFINITIONS.

236. To represent a body with accuracy, it is not only necessary that the drawing should give the representations of the details of its form, but also of its colors, whether natural or artificial, or the effect of light and shade.

Different portions of the same body will appear lighter or darker according as the light falls directly upon it or is excluded from it, by itself or some other body. A simple application of the elementary principles previously deduced will enable us to limit and represent these portions, and constitutes that part of the subject called *Shades* and *Shadows.*

237. It is a principle of Optics, that the effect of light, when in the same medium and unobstructed, is transmitted from each point of a luminous body in every direction, along right lines. These right lines are called *rays of light. Any right line*, therefore, *drawn from a point of a luminous body* will be regarded as a ray of light.

The sun is the luminous body which is the principal source of light. It is at so great a distance, that rays drawn from any of its points to an object on the earth's surface, may, without material error, be regarded as parallel. In the construction of problems, in this part of our subject, they will be so taken.

238. Let SR, Fig. 103, indicate the direction of the parallel rays of light, coming from the source and falling on the opaque body B; and let *n-mlop-q* be a cylinder, whose rectilinear elements are rays of light parallel to SR, enveloping and touching this body.

It is evident that all light, coming directly from the source, will be excluded from that part of the interior of this cylinder which is behind B. This *part of space from which the light is* thus *excluded by the opaque body, is the indefinite shadow of the body;* and any object within this portion of the cylinder is *in the shadow*, or *has a shadow cast upon it.*

The line of contact *mlop* separates the surface of the opaque body into two parts. That part which is towards the source of light, and on which the rays fall directly, is *the illuminated part;* and that part opposite the source of light from which the rays are excluded by the body itself, is *the shade of the body.*

This *line of contact* of the tangent cylinder of rays, thus *separating the illuminated part from the shade*, is *the line of shade.* Any plane tangent to this cylinder of rays will also be tangent to the opaque body at some point of this line of contact, Art. (116); and, conversely, every plane tangent to the opaque body at a point of the line of shade will be tangent to this cylinder, and contain a rectilinear element, Art. (110), or ray of light, and thus be *a plane of rays.* Hence points of the line of shade on any opaque body may be obtained *by passing planes of rays tangent to the body, and finding their points of contact.*

239. If the cylinder *n-mlop-q* be intersected by any surface, as the plane ABCD, that part *m'l'o'p'* of this plane, which is in the shadow, is the shadow of B *on* this plane. That is, *the shadow of an opaque body on a surface is that part of the surface* from

which the rays of light are excluded by the interposition of this body between it and the source of light.

The bounding line of this shadow, or *the line which separates* the shadow on a surface from the illuminated part of that surface, is *the line of shadow.* It is also the line of intersection of the cylinder of rays which envelops the opaque body, and the surface on which the shadow is cast.

240. When the opaque body is bounded by planes, the cylinder of rays, by which its shadow is determined, will be changed into several *planes of rays*, which will include the indefinite shadow; and the line of shade will be made up of right lines which, though not lines of mathematical tangency, are the outer lines in which these planes touch the opaque body, and still *separate the illuminated part from the shade.*

The line of shadow, in this case, is also made up of the lines of intersection of these planes with the surface on which the shadow is cast, and still separates the illuminated part of this surface from the shadow.

SHADOW OF POINTS AND LINES.

241. The indefinite shadow of a point may be regarded as that part of a ray of light, drawn through it, which lies in the direction opposite to that of the source of light, and the point in which this ray pierces any surface is *the shadow* of the point *on that surface.*

242. The shadow of a right line will then be determined by drawing through its different points rays of light. These form *a plane of rays*, and the intersection of this plane with any surface is *the shadow of the line on that surface.*

To determine whether a given plane is a plane of rays, it is only necessary to ascertain whether *it contains a ray of light.* This may be done by drawing through any one of its points a ray. If this pierces the planes of projection in the traces of the given plane, it is a plane of rays, Art. (30).

243. *The shadow of a right line on a plane* may be constructed by finding the shadows cast by any two of its points on the plane, Art. (241), and joining these by a right line; or by joining the shadow of any one of its points with the point in which the line pierces the plane, Art. (30).

If the right line *be parallel to the plane,* its shadow on the plane will be parallel to the line itself, and the shadow of a definite portion of it will be equal to this portion, Art. (14).

If the line is a ray of light, its shadow on any surface will be a point.

Thus let MN, Fig. 104, be a limited right line, and tTt' the plane on which its shadow is required, MR indicating the direction of the rays of light.

Through the extremities M and N, draw the rays MR and NS. They pierce tTt' in R and S, Art. (40), which are points of the shadow, Art. (241). Join these points by RS, it will be the required shadow.

If the shadow be cast on the horizontal plane, the rays through the extremities M and N, Fig. (*a*), pierce H at m'' and n'', and $m''n''$ is the shadow.

If the shadow be cast on the vertical plane, these rays pierce V at m'' and n'', Fig. (*b*), and $m''n''$ is the shadow.

If MN is parallel to either plane, then RS in Fig. 104, and $m''n''$ in Figs. (*a*) and (*b*) will be equal to MN.

If right lines are parallel, their shadows on a plane are parallel, since they are the intersections of parallel planes of rays by this plane.

244. The shadow of a curve will be determined by passing through it a cylinder of rays. *The intersection of this cylinder with any surface* will be the shadow of the curve *on* that surface.

The line of shadow on a surface, Art. (239), will always be the shadow of the line of shade; and if through any point of the line of shadow a ray be drawn to the source of light, it will intersect the line of shade in a point which casts the assumed point of shadow.

From this it follows that *the point of shadow cast by any line in space upon any other line*, may be found by constructing the shadows of both lines on a plane; and drawing through the point of intersection of these shadows a ray, it will intersect the second line in the point of shadow cast upon it, by the point in which it intersects the first.

245. If *two lines are tangent* in space, their shadows, on any surface, will be tangent at the point of shadow cast by the given point of contact. For the cylinders of rays through the lines will be tangent along the ray passing through the point of contact, Art. (117); hence their intersections by any surface will be tangent at the point where this ray pierces the surface. These intersections are the shadows of the lines.

246. The *shadow of a curve of single curvature*, on a plane to which it is parallel, will be an equal curve, since each of its elements will cast a parallel and equal element of the shadow, Art. (243).

If the plane of the curve be a plane of rays, its shadow on a plane will evidently be a right line.

The *shadow of the circumference of a circle*, on a plane to which it is parallel, will be *an equal circumference*, whose centre is the shadow cast by the given centre.

Also, when the plane on which the shadow is cast makes a *sub-contrary section* in the cylinder of rays through the circumference, Art. (206).

In all other positions, when its plane is not a plane of rays, its shadow will be an ellipse, Art. (160), and any two diameters of the circle *perpendicular to each other* will cast *conjugate diameters of the ellipse* of shadow. For if at the extremities of either diameter tangents be drawn, they will be parallel to the other diameter, and their shadow parallel to the shadow of this diameter, Art. (243). But the shadows of the two tangents are tangent to the shadows of the circle, at the extremities of the shadow of the first diameter, Art. (245); hence the shadow of the second diameter is parallel to the tangents at the extremities of the shadow of the first. These shadows are therefore conjugate diameters of the ellipse, and with them the ellipse may be constructed. *Ana. Geo.*, Art. (150).

If the shadows of two perpendicular diameters be found, also perpendicular to each other, they will be the axes of the ellipse, which may be constructed as in Art. (59).

BRILLIANT POINTS.

247. In looking upon the illuminated part of a curved surface, it will be observed that, in general, one or more points appear much more brilliant than the others. This is due to the fact that *the ray of light, incident at each of these points, is reflected immediately to the point of sight.*

These points are called *brilliant points;* and since it is a principle of Optics that the incident and reflected rays, at any point of a surface, lie in the same normal plane on opposite sides of the normal at this point, and making equal angles with it, it follows that at a brilliant point, the ray of light and the right line drawn to the point of sight must fulfil these conditions.

When the point of sight is at an infinite distance, as in the

Orthographic projection, the brilliant point may be constructed thus: Through any point draw a ray of light and a right line to the point of sight. These will be parallel, respectively, to the corresponding lines at the required point, and make the same angle. Bisect the angle formed by these auxiliary lines as in Art. (37). The bisecting line will be parallel to the line which bisects the parallel and equal angle at the brilliant point. Hence, *perpendicular to this bisecting line, and tangent to the surface*, pass a plane. Its point of contact *will be the brilliant point;* for a line through this point parallel to the bisecting line will be a normal to the surface, and will bisect the angle formed by a ray of light and right line drawn from this point to the point of sight.

PRACTICAL PROBLEMS.

248. PROBLEM 61. *To construct the shadow of a rectangular pillar, on the planes of projection.* Let *mnlo*, Fig. 105, be the horizontal and *m'n'p'q'* the vertical projection of the pillar, and let (*nn''*, *n'r*) indicate the direction of the rays of light.

The upper base vertically projected in *m'n'*, and the two vertical faces horizontally projected in *mn* and *mo*, are evidently illuminated, and together form the illuminated part of the pillar; and the edges (*n*, *p'n'*), (*nl*, *n'*), (*lo*, *n'm'*), and (*o*, *q'm'*) separate the illuminated part from the shade, and make up the line of shade, Art. (240).

Since *n* is in the horizontal plane and *n''* the shadow of N, *nn''* is the shadow of (*n*, *p'n'*) on H, Art. (243). The plane of rays through (*nl*, *n'*) is perpendicular to V, and *n''r*, parallel to *nl*, is its horizontal and *rl''* its vertical trace; hence *n''r* is that part of the shadow of (*nl*, *n'*) which is on H; and *rl''*, limited by the ray through (*ln'*), that part on V. The ray (*rx*, *rn'*) intersects (*nl*, *n'*) in (*xn'*), and *nx* is the horizontal projection of that part which casts the equal shadow *n''r*, Art. (243), and *xl* of that part which casts *rl''*.

$l''o''$ is the shadow of $(ol, m'n')$ on V, parallel and equal to itself.

The plane of rays through $(o, q'm')$ is perpendicular to H, and os is its horizontal and so'' its vertical trace; and os the shadow of $(o, q'm')$ on H, and so'' its shadow on V, $q'y'$ is the vertical protection of the part which casts the shadow os and $y'm'$ of that part which casts its equal so'', Art. (243).

The line of shadow is thus the broken line $nn''r...o''...o$, and the portion of the planes within this line is the required shadow, and should be darkened as in the drawing.

249. Problem 62. *To construct the shadow of a rectangular abacus on the faces of a rectangular pillar.*

Let $mnlo$, Fig. 106, be the horizontal, and $m'n'l'o'$ the vertical projection of the abacus, and $cdd'c'$ the horizontal, and $c'd'e'f'$ the vertical projection of the pillar; MR indicating the direction of the rays.

The lines of shade on the abacus, which cast the required shadows, are evidently the two edges (mo, m') and MN.

The plane of rays through (mo, m') is perpendicular to V, and intersects the side face, horizontally projected in cc', in a right line perpendicular to V at s', which is the shadow on this face. $m'r'$ is the vertical trace of this plane. The ray MR, through M, pierces the front face of the pillar in R, Art. (40), the shadow of M, and $(cr, s'r')$ is the shadow of the part (pm, m') on this face.

MN being parallel to this face, its shadow on it will be parallel to itself, and pass through R; hence $(rd, r'x')$ is the required shadow cast by its equal MQ.

250. Problem 63. *To construct the shadow of an upright cross upon the horizontal plane and upon itself.*

Let $mnop$, Fig. 107, be the horizontal, and $c'n'r'g'f'$ the vertical projection of the cross.

cc'' is the shadow of the edge $(c, c'd')$, c'' being the shadow of

(cd'); $c''n''$ is the shadow of its equal $(cn, d'n')$, Art. (243); $n''h''$ of the edge $(n, n'h')$; $h''o''$ of its equal (no, h'); $o''y''$ of $(oy, h'x')$; (xy, x') of a part of (cd, r'), on the face of the cross vertically projected in $l'h'$. $y''r''$ is the shadow of the remaining part of (cd, r') on H. $(cx, l'x')$ is the shadow of $(c, l'r')$ on the face vertically pro-ected in $l'h'$, this being the intersection of a vertical plane of rays through $(c, l'r')$ with this face. $r''q''$ is the shadow of its equal $(de, r'q')$; $q''k''$ of $(e, k'q')$; $k''p''$ of $(pe, k'g')$; $p''m''$ of $(p, m'g')$; and $m''t$ of a part of (pm, m').

All within the broken line thus determined on H is darkened, as also the part *cxyd*, the horizontal projection of the shadow cast by the upper part of the cross on the face vertically projected in $l'h'$.

251. Problem 64. *To construct the shade of a cylindrical column, and of its cylindrical abacus, and the shadow of the abacus on the column.*

Let *mlo*, Fig. 108, be the horizontal, and $m'o'h'n'$ the vertical projection of the abacus, *cde* the horizontal, and $cee'g'$ the vertical projection of the column.

Pass two planes of rays tangent to the column. Since each contains a rectilinear element, they will be perpendicular to H; and *ld* and *kf*, parallel to the horizontal projection of the ray of light, will be their horizontal traces, Art. (123); and $(d, z'd')$ and $(f, f'r')$ the elements of contact, which are indefinite lines, *or elements of shade*, Art. (238.)

In the same way the elements of shade $(i, u'i')$, &c., on the abacus, are determined.

The line of shadow on the column will be cast by a portion of the lower circumference of the abacus towards the source of light. To determine any point of it, pass a vertical plane of rays, as that whose horizontal trace is *yx*. It intersects the column in a rectilinear element $(x, x''x')$, and the circumference in a point Y. Through this point draw the ray YX. It intersects the element in X, a point of the required shadow, Art. (241) · and in the same way any number of points may be found. C is the shadow of Q, and D of L, and $c'x'd'$ is the vertical, and cxd the horizontal projection of that part of the curve of shadow which can be seen.

That part of the shade on the abacus and column, and of the shadow on the column, which can be seen, is darkened in the drawing.

252. PROBLEM 65. *To construct the shade of a right cylinder with a circular base, and its shadow on the horizontal plane and on its interior surface.*

Let mlo, Fig. 109, be the base of the cylinder, and $(c, d'c')$ its axis.

The two planes of rays, whose horizontal traces are ll'' and kk'', determine the two *elements of shade* $(l, n'l')$ and $(k, r'k')$.

These elements cast the shadows ll'' and kk'' on H, Art. (243).

The semi-circumference of the upper base LUK casts its shadow on the interior of the cylinder, and the other semi-circumference on the horizontal plane, without the cylinder.

The shadow of the latter is the equal semi-circumference $k''o''l''$ whose centre is c'', Art. (246), and this, with the lines ll'' and kk'', limits the shadow of the cylinder on H.

To determine the shadow on the interior surface, pass any vertical plane of rays, as that whose horizontal trace is yx. It intersects the cylinder in the element $(x, x''x')$ and the semi-circumference in the point Y. The ray of light YX intersects $(x, x''x')$ in X, a point of the required shadow; and in the same way any number of points may be determined.

The shadow upon any rectilinear element, may be found by using an auxiliary plane which shall contain this element. Thus Z is the shadow of U on the element $(z, z''z')$, s'' of S, and t'' of T.

This shadow evidently begins at L and K, the upper ends of the elements of shade.

The direction of the rays is so taken that a part of this semi-circumference casts its shadow on H, within the cylinder. This part is horizontally projected in *st*, and its shadow is the equal arc $s''t''$ the continuation of $k''o''l''$.

Should this shadow not reach H, *its lowest point* will be obtained by using as an auxiliary plane that one which contains the axis; and the point in which the vertical projection of the curve of shadow is tangent to $o'p'$, by using the plane which contains the element $(o, o'p')$.

That part of the shade between the elements $(l, n'l')$ and $(o, o'p')$ only, can be seen when looking on V, and is darkened in the drawing.

253. Problem 66. *To construct the shade of an oblique cone and its shadow on the horizontal plane.*

Let the cone be given as in Fig. 110, S being its vertex.

If two planes of rays be passed tangent to the cone, the elements of contact will be the *elements of shade*, Art. (238). Since these planes must pass through S, Art. (130), they must contain the ray of light SR. This pierces H at *r*, and *rn* and *rp* are the horizontal traces of these tangent planes, and SN and SP are the elements of contact, Art. (130).

These elements cast the shadows *nr* and *pr*, which limit the shadow of the cone on H.

In looking upon H, the shade between the elements SP and SQ only is seen; and in looking upon V, that between SN and SO.

254. Problem 67. *To construct the shade of an ellipsoid of revolution and its shadow on the horizontal plane.*

Let the ellipsoid be given as in Fig. 111.

The line of shade is the line of contact of a cylinder of rays tangent to the surface, and is *an ellipse, Ana. Geo.*, Art. (233); and since the meridian plane of rays, whose horizontal trace is *cs*, is evidently a principal plane, Art. (206), of both the ellipsoid and cylinder, bisecting all chords of both surfaces perpendicular to it, its intersection with the plane of the ellipse will also bisect all chords of the ellipse perpendicular to it, and be an axis of the curve.

This meridian plane cuts from the ellipsoid a meridian curve, and from the cylinder two rectilinear elements, rays of light, tangent to this curve, at the vertices of the axis, *the highest and lowest points* of the curve of shade.

To determine these points, revolve the meridian plane about (c, $e'd'$), until it becomes parallel to V. The meridian curve will then be vertically projected into $e'm'd'n'$; CT will be the revolved position of CS, the axis of the cylinder, and the two tangents at r' and x', parallel to $c't'$, will be the vertical projections, in revolved position, of the two elements cut from the cylinder, and R and X will be the revolved position of the two vertices. When the plane is revolved to its true position, these points will be at P and Y, Art. (107), and PY is the transverse axis of the required curve.

The conjugate axis, being perpendicular to the meridian plane of rays, is parallel to H, and therefore horizontally projected into fg, perpendicular to yp, Art. (36), and vertically into $f'g'$.

The projections of this curve are ellipses. Of the horizontal projection, fg is evidently the longest diameter, and therefore the transverse axis, and yp the conjugate, and the ellipse $fpgy$ is the horizontal projection of the curve of shade.

Since the tangents to the curve at the highest and lowest points P and Y are horizontal, their vertical projections must be $r'p'$ and $x'y'$ tangent to the vertical projection of the curve at p' and y'. These tangents being parallel to $f'g'$, $p'y'$ and $f'g'$ must be conjugate diameters of the ellipse which is the vertical projection of the curve; hence $p'f'y'g'$, on these diameters, is the required vertical projection.

Points of this curve may also be found otherwise, by intersecting the surface by any horizontal plane, as that of which $l'k'$ is the vertical trace. It cuts from the surface a circumference, and from the plane of the curve a line parallel to FG. These lines intersect in the required points. K and L are thus determined.

The points of tangency, u' and w', are the vertical projections of the points in which the curve of shade is intersected by the meridian plane parallel to V.

The shadow cast by the curve of shade on H is an ellipse, the intersection of the tangent cylinder with H. The conjugate axis of this ellipse, $f''g''$, is the shadow of FG parallel and equal to itself, and the semi-transverse axis, $y''s$, the shadow of CY, since these shadows are perpendicular to each other, and $y''s$ bisects all chords perpendicular to it.

255. If the ellipsoid becomes a sphere, the construction of the curve of shade is simplified, as it then becomes the circumference of a great circle perpendicular to CS, Art. (121), and fg is the projection of that diameter which is parallel to H, and py the projection of the one which is perpendicular to it, Art. (197), and the vertical projection may be obtained in the same way.

256. If a plane of rays be passed tangent to any surface of revolution, the point of contact will be a point of the curve of shade, Art. (238). If through this point of contact a meridian plane be passed, it will be perpendicular to the tangent plane, Art. (115), and cut from it a line tangent to the meridian curve at this point, Art. (108). The plane, which projects a ray of light on to this meridian plane, is also a plane of rays perpendicular to it, and therefore parallel to the tangent plane, and its intersection with the meridian plane will be parallel to the tangent cut from the tangent plane. But this intersection is the projection of the ray. Hence, *if a tangent be drawn to any meridian curve of a surface of revolution parallel to the projection of a ray of light on the meridian plane, its point of contact will be a point of the line of* *shade*, and thus any number of points may be found on different meridians, and the line of shade drawn.

The points U and W, in Fig. 111, may thus be determined by drawing tangents to the meridian curve which is parallel to V, parallel to $c's'$.

257. PROBLEM 68. *To find the brilliant point on any surface of revolution.*

Let the surface be an ellipsoid, given as in Fig. 111. Through any point of the axis as C, draw the ray of light CS and the line (co, c') to the point of sight in front of the vertical plane. Bisect their included angle as in Art. (37) by the line CQ. To determine a plane perpendicular to this line and tangent to the surface, Art. (247), through it pass the meridian plane of which cq is the horizontal trace. This plane cuts from the surface a meridian curve, and from the required tangent plane a right line tangent to this curve at the required point, and perpendicular to CQ.

Revolve this plane about (c, $c'd'$) until it becomes parallel to V. The meridian curve will be vertically projected in $e'n'd'm'$, and CQ in $c'q'''$, Art. (29). Tangent to $e'm'd'$ and perpendicular to $c'q'''$ draw $i'z'''$; z''' is the vertical projection of the revolved position of the required point, horizontally projected at z'', and when the plane is revolved to its true position at z. Z is then the required brilliant point.

258. PROBLEM 69. *To construct the shade and shadow of an upright screw.*

Let ecf, Fig. 112, be the circular base of a right cylinder whose axis is oc', and let $em'e'$ be an isosceles triangle in the plane of the element ee' and of the axis, its base coincident with the element. If this triangle be moved about the cylinder, its plane always containing the axis, with a uniform angular motion, and at the same time with a uniform motion in the direction of the axis, so that after passing around once it will occupy the position $e'm''e''$, it will generate a volume called *the thread of a screw.*

The cylinder is *the cylinder of the screw.*

It is evident that the two sides $m'e$ and $m'e'$ will each generate a portion of a helicoid, Art. (92), the side $m'e'$ generating *the upper surface* of the thread, and $m'e$ *the lower surface.*

The point m' generates *the outer helix* of the thread, and the point e *the inner helix.*

The curve of shade on the lower surface of the thread may be constructed by passing planes of rays tangent to this surface, and joining the points of contact by a line, Art. (238).

To find the point of this curve *on the inner helix*, pass a plane tangent to the lower helicoid at F; gd is the horizontal trace of this plane, the distance fd being equal to the rectified arc fce, Art. (139).

It is a property of the helicoid that all tangent planes to the surface, *at points of the same helix*, make the same angle with the axis, or with the horizontal plane, when taken perpendicular to this axis. For each of these planes is determined by a tangent to the helix, and the rectilinear element through the point of contact; and these lines at all points of the helix make the same angle with the axis, Arts. (69) and (92), and with each other.

Hence, if through the axis we pass a plane perpendicular to the tangent plane at F, it will cut from this plane a right line, intersecting oc' at o'', horizontally projected in oi, and making with H the same angle as that made by a plane tangent at any point of the helix, and with the rectilinear element passing through the point of contact an angle, which is the same for all these planes; hence the angle iof will be the same for all these planes.

If this line be now revolved about oc', it will generate a cone whose rectilinear elements make the same angle with H as the tangent plane. If a plane of rays be passed tangent to this cone, it will be parallel to the plane of rays tangent to the surface at a point on the inner helix. kl, tangent to the circle generated by oi, is the horizontal trace of this plane, Art. (131), and ol is the horizontal projection of the line cut from the parallel tangent plane of rays by a perpendicular plane through *oc'*. If, then, we make the angle *lox* equal to *iof*, *ox* will be the horizontal projection of the element containing the required point of contact, and X will be the point.

In the same way, the point Y on the outer helix may be obtained, by first passing a plane tangent to the lower surface at M. *eq* is its horizontal trace, *mq* being equal to the rectified semi-circumference *myn*, and *eop* the constant angle. A tangent from *t* to the circle generated by *op* (so nearly coincident with *ecf* that it is not drawn) is the horizontal trace of the parallel plane of rays, and *oy*, making with the right line joining *o* and the point of contact of this tangent, an angle equal to *eop*, the horizontal projection of the element containing the point of contact, and Y the required point.

Intermediate points of the curve of shade may be determined by constructing, as above, the points of contact of tangent planes of rays on intermediate helices. Thus, pass a plane tangent to the lower helicoid at W, midway between F and N; *wbz* is the horizontal projection of the helix passing through this point; *fb''* the horizontal projection of the intersection of the tangent plane with the horizontal plane through *f'*, *wb''* being equal to the rectified arc *wb*. Since this intersection is parallel to the horizontal trace of the tangent plane, *ou* will be the horizontal projection of the line cut from the tangent plane by the perpendicular plane through *oc'*, and *fou* the constant angle. *us* is the horizontal projection of the circle cut from the auxiliary cone by the horizontal plane through *f'm'*, and *ts* the horizontal projection of the intersection of the tangent plane of rays with the same plane; *oz* making with the right line joining *o* and *s*, an angle equal to *fou*, is then the horizontal projection of the element containing the point of contact, *o'''v'* Art. (92) its vertical projection, and Z the required point.

As the surface, of which *e'm''n''f''* is the vertical projection, is in all respects identical with that of which *em'n'f'* is the projection, $y''x''$ will be the vertical projection of the curve of shade on the first surface, its horizontal projection being yx.

To construct *the shadow cast by this curve* of shade on the surface of the thread below it, construct the shadow of $(xy, x''y'')$ on the horizontal plane through $n'e'$, under the supposition that the rays are unobstructed. $x'''y'''$ is the horizontal projection of this shadow, Art. (244.) Then assume any element, as $(\alpha o, \alpha' o_{,})$, on which it is supposed the shadow will fall, and construct its shadow on the same plane. $h''r''$ is the horizontal projection of this shadow. Through the point in which these shadows intersect, horizontally projected at r'', draw a ray of light, intersecting the element in R, which is the shadow of the curve on the element, Art. (244.) The curve evidently begins at (x, x'') and is vertically projected in $x''r'$.

The shadow cast by the helix $(mn, m''n'')$ on the surface of the thread may be constructed in the same way by first constructing its shadow on the same horizontal plane as above, and the shadow of an assumed element $(o\delta, o_2\delta')$ on the same plane, and from their point of intersection horizontally projected at δ'', drawing a ray; (δ, δ') will be a point of the required shadow.

The curve of shadow on the surface, whose vertical projection is enf', will be equal to the curve whose vertical projection is $x''r'\delta'$, and may be drawn as in the figure.

259. *The brilliant point* on the upper surface of the thread may be constructed by bisecting the angle formed by a ray of light and a line drawn to the point of sight, Art. (247), and then passing a plane perpendicular to the bisecting line and tangent to the surface, as in Art. (140), and finding its point of contact.

PART IV.

LINEAR PERSPECTIVE.

PRELIMINARY DEFINITIONS AND PRINCIPLES.

260. It has been observed, Art. (204), that the orthographic projections can never present to the eye of an observer a perfectly natural appearance, and hence this mode of representation is used only in drawings made for the development of the principles, and the solution of problems in Descriptive Geometry, and for the purposes of mechanical or architectural constructions.

Whenever an accurate picture of an object is desired, the scenographic method must be used, and the position of the point of sight chosen, as indicated in Art. (204).

261. That application of the principles of Descriptive Geometry which has for its object *the accurate representation, upon a single plane, of the details of the form and the principal lines of a body*, is called *Linear Perspective.*

The art by which a proper coloring is given to all parts of the representation, is called *Aerial Perspective.* This, properly, forms no part of a mathematical treatise, and is therefore left entirely to the taste and skill of the artist.

262. The plane upon which the representation of the body is made is called *the plane of the picture;* and a point is represented upon it, as in all other cases, Art. (3), by drawing through the point and the point of sight a right line. The point in which it pierces the plane of the picture is *the perspective of the given point.*

These projecting lines are called *visual rays*, and when drawn to the points of any curve, form *a visual cone*, Art. (77).

Any plane, passing through the point of sight, is made up of visual rays, and is called *a visual plane.*

The plane of the picture is usually taken between the object to be represented and the point of sight, in order that the drawing may be of smaller dimensions than the object. It is also taken vertical, as in this position it will, in general, be parallel to many principal lines of the object. It may thus be used as the vertical plane of projection, and will be referred to as the plane V, Art. (24).

The orthographic projection of the point of sight, on the plane of the picture, is called *the principal point of the picture;* and a horizontal line through this point and in the plane of the picture, is *the horizon of the picture.*

PERSPECTIVES OF POINTS AND RIGHT LINES. VANISHING POINTS OF RIGHT LINES.

263. Let M, Fig. 113, be any point in space, AB the horizontal trace of the plane of the picture, or ground line, and S the point of sight.

The visual ray SM, through M, pierces the plane of the picture V, in m_1 Art. (40), which is the perspective of the point M.

264. The indefinite perspective of a right line, as MN, Fig. 114, may be drawn by finding the perspectives of any two of its points

as in the preceding article, and joining them by a right line; or by passing through the line a visual plane, and constructing its trace on the plane of the picture.

The point n', in which this line pierces V, is one point of this trace, Art. (30). If through S a line SP be drawn parallel to MN, it will lie in the visual plane, and pierce V at p', a second point of the trace, and $p'n'$ will be the perspective of the line.

The point p' is called *the vanishing point* of the line MN. Hence, to construct the perspective of any right line, *join its vanishing point with the point where the line pierces the plane of the picture, by a right line.*

265. *The vanishing point* of any right line is *the point in which a line, parallel to it through the point of sight, pierces the plane of the picture.*

This point, as seen in the preceding article, is always one point of the perspective of the line; and since the auxiliary line intersects the given line only at an infinite distance, this point is the perspective of that point of the given line which is at an infinite distance.

The principal point is evidently the vanishing point of *all perpendiculars* to the plane of the picture.

266. *Any horizontal right line* making an angle of 45° with the plane of the picture is *a diagonal.*

A right line through S, Fig. 113, parallel to a diagonal, pierces V in the horizon, and at a distance from the principal point s', equal to the distance of S from the plane of the picture. Since two such lines can be drawn, one on each side of the perpendicular Ss', there will be two vanishing points of diagonals, one for those which incline to the right, and another for those which incline to the left.

These points are also called *points of distance*, since when assumed or fixed, the distance of the point of sight from the plane of the picture is determined.

267. If *a plane be passed through the point of sight parallel to a given plane*, it will contain all right lines drawn through this point parallel to lines of the plane. Its trace on the plane of the picture will therefore contain the vanishing points of all these lines, and of lines parallel to them. This trace is *the vanishing line of the plane.*

The horizon is evidently the vanishing line of all horizontal planes.

A system of parallel right lines will have a *common vanishing point;* since the line through the point of sight parallel to one is parallel to all; hence their perspectives all intersect at this point, Art. (265).

If the lines are also *parallel to the plane of the picture*, the auxiliary line will be parallel to it, and their vanishing point at an infinite distance. Their perspectives will then be parallel, and *the perspective of each line parallel* to itself, and it will be necessary to determine only one point of its perspective.

If a right line pass through the point of sight, it is a visual ray, and *the point* in which it pierces the plane of the picture is its perspective.

268. If a point be on a line, its perspective will be on the perspective of the line. If then through any point two right lines be drawn, and their perspectives found as in Art. (264), the intersection of these perspectives will be the perspective of the given point. Hence, in practice, the perspective of a point is constructed *by drawing the perspectives of a diagonal and perpendicular*, which, in space, pass through the point, and *finding the intersection of these perspectives.*

Thus let s', Fig. 113, be *the principal point*, $s'd_1$ the horizon, d_1 one point of distance, and M any point in space.

MN is a diagonal through the point its vertical projection $m'n$ being parallel to AB, Art. (14). It pierces V at n', vanishes at d_1 and $n'd_1$ is its perspective, Art. (264). The perpendicular through M pierces V at m', vanishes at s', and $m's'$ is its perspective. These perspectives intersect at m_1 the perspective of M.

If the point M should be in the horizontal plane, the diagonal and perpendicular would pierce V in AB.

When the given point is near the horizontal plane through the point of sight, the perspectives of the diagonal and perpendicular through it are so nearly parallel that it is difficult to mark accurately their point of intersection. In this case, find the perspective of a vertical line through the given point; its intersection with the perspective of the diagonal or perpendicular will be the required point. Thus in Fig. 113, find m_2, the perspective of m the foot of a vertical line through M, and draw m_2m_1 intersecting $m's'$ in m_1 the perspective of M.

269. Since the object to be represented is usually behind the plane of the picture, Art. (262), and is given by its orthographic projections, these projections when made as in Art. (4), will occupy the same part of the drawing as the perspective, and cause confusion. To avoid this, in some degree, that portion of the horizontal plane which is occupied by the horizontal projection of the object, is revolved about AB 180°, until it comes in front of the plane of the picture. Thus, in Fig. 113, m comes to m'', and the horizontal projection of the diagonal MN to the position $m''n$, its vertical projection being in its primitive position, and the point n' being found as before, Art. (27). It will be observed that the horizontal projection of each diagonal, in this new position, lies in a direction contrary to that of its true position. The vanishing point will be determined by its true direction, Art. (266).

PERSPECTIVES OF CURVES.

270. The perspective of any curve may be found by joining its points with the point of sight by visual rays, thus forming a visual cone. The intersection of this cone by the plane of the picture will be the required perspective: Or the perspectives of its points may be found as in Art. (268), and joined by a line.

If the curve be of single curvature, Art. (58), and its plane pass through the point of sight, its perspective will be *a right line.*

271. *If two lines are tangent in space, their perspectives will be tangent.* For the visual cones, by which their perspectives are determined, will be tangent along the common rectilinear element passing through the point of contact of the lines; hence their intersection by the plane of the picture will be tangent at the perspective of this point of contact.

272. *If the circumference of a circle* be parallel to the plane of the picture, its perspective will be the circumference of a circle whose centre is the perspective of the given centre; also, if it be so situated that the plane of the picture makes in its visual cone a sub-contrary section, Art. (206).

In all other cases, when behind the plane of the picture, and its plane does not pass through the point of sight, its perspective will be *an ellipse.*

It is evident that if the condition be not imposed that the circle shall be behind the plane of the picture, this plane may be so taken as to intersect the visual cone in *any of the conic sections.*

273. Let *mlo*, Fig. 115, be a circle in the horizontal plane, revolved as in Art. (269), and let s' be the principal point, the

point of sight being taken in a plane through the centre and perpendicular to AB, and let d_1 and d_2 be *the points of distance.*

o_1 is the perspective of o, and m_1 of m, Art. (268), and o_1m_1 of the diameter om. The perspectives of the two tangents at o and m will be tangent to the perspective of the circle at o_1 and m_1, Art. (271); and since the tangents are parallel to AB, their perspectives will be parallel to AB, Art. (267), or perpendicular to o_1m_1; hence o_1m_1 is *an axis* of the ellipse. Through its middle point e_1 draw d_1e_1; it is the perspective of the diagonal fe, and e_1 is the perspective of e, and u_1z_1 of the parallel chord uz, $u's'$ and z_1s' being the perspectives of the perpendiculars through u and z, and u_1z_1 is the other axis of the ellipse. On these two axes the ellipse can be described as in Art. (59).

The perspectives of the two perpendiculars ll' and kk' are tangent to the ellipse at l_1 and k_1.

If the point of sight be taken in any other position, the perspective of the circle, Fig. 116, may be determined by points, as in Art. (270), and the curve drawn through these points tangent to the perspectives of the tangents at o, m, l and k. The line o_1m_1 will be a conjugate diameter to the line u_1z_1, since the tangents at o_1 and m_1 are parallel to AB and to u_1z_1.

LINE OF APPARENT CONTOUR.

274. If a body, bounded by a curved surface, be enveloped by a tangent visual cone, the line of contact of this cone will be the outer line of the body, as seen from the point of sight, and is *the line of apparent contour* of the body.

The perspective of this line, Art. (270), is evidently the bounding line of the picture or drawing, and is a principal line of the perspective.

When the body is bounded by plane surfaces, the visual cone will be made up of visual planes; the line of apparent contour will not be a line of mathematical contact, but will still be the outer line of the body as seen, and will be made up of right lines.

When the body is irregular, or composed of broken surfaces, the line of contour may be composed partly of strict lines of contact, either straight or curved, and partly of lines not of contact, but still the outer lines of the body as seen. In all cases their perspectives will form the boundary of the picture.

To *construct the perspective of any body*, we must then determine the perspective of its apparent contour, and of such other principal lines as will aid in indicating its true form.

275. If a line of the surface *intersect the line of contact* of the visual cone, *the perspective of this line* will be tangent to the perspective of the line of contact. For at the point of intersection draw a tangent to each of the two lines; these tangents lie in the tangent plane to the surface, and this plane is also a visual plane tangent to the cone. Its trace on the plane of the picture is the perspective of both tangents, and tangent to the perspectives of both lines at a common point. They are therefore tangent to each other.

VANISHING POINTS OF RAYS OF LIGHT AND OF PROJECTIONS OF RAYS.

276. Since the rays of light are parallel they will have a common vanishing point, which may always be determined by drawing a visual ray of light, and finding the point in which it pierces V, Art. (265). In the construction of a drawing or picture, the direction of the light is usually chosen by the draughtsman or artist. This is done by assuming the vanishing point of rays at once, as the direction of the light is thus completely determined. Thus let r_1, Fig. 117, be the vanishing point of rays, S being the point of sight, sr will be the horizontal projection of the ray, and $s'r_1$ its vertical projection.

277. The horizontal projections of rays of light being parallel and horizontal lines, must vanish in the horizon, Art. (267); and since a line through the point of sight parallel to the horizontal projection of a ray, must be in the same vertical plane with a ray through the same point, it must pierce the plane of the picture in the vertical trace of this plane, that is, in the line r_1r' at r'. Hence, having assumed the vanishing point of rays, through it draw a right line perpendicular to the horizon; it will intersect it in *the vanishing point of the horizontal projections of rays.*

278. The orthographic projections of rays on all planes perpendicular to AB, as the plane tTt', are parallel. The line $s's$ is the vanishing line of these planes, Art. (267), and must therefore contain the vanishing point of these projections. This point must also be in the trace of a plane of rays through S, perpendicular to these side planes. r_1r_2, parallel to AB, is this trace; hence r_2 is *the vanishing point of the projections of rays on side planes.*

279. The orthographic *projections of rays on the plane of the picture*, or *on planes parallel to it*, are parallel; and being in, or parallel to, the plane of the picture, will be parallel in perspective, Art. (267), and all parallel to r_1s', the projection of the ray through S on V.

PERSPECTIVES OF THE SHADOWS OF POINTS AND RIGHT LINES ON PLANES.

280. Since *the shadow of a point on a plane* is the point in which a ray of light through the point pierces the plane, Art. (241), it must be the intersection of the ray with its orthographic projection on the plane. Hence, to construct *the perspective of the shadow* of a point on any plane, through the perspective of the point draw

the perspective of a ray, and through the perspective of the projection of the point on the plane, draw the perspective of the projection of the ray; the intersection of these two lines will be the perspective of the required shadow, Art. (268).

If *the shadow be on the horizontal plane*, join the perspective of the point with the vanishing point of rays r_1, Fig. 117, and the perspective of the horizontal projection of the point with the vanishing point of horizontal projections r'; the intersection of these two lines will be the perspective of the shadow of the given point.

If *the shadow be on any side plane*, join the perspective of the projection of the point on this plane with the vanishing point of the projections of rays on side planes r_2; the intersection of this line with the perspective of the ray, will be the perspective of the shadow.

If *the shadow be on any plane parallel to* V, through the perspective of the projection of the point on this plane draw a line parallel to the projection of the ray on the plane of the picture $s'r_1$; its intersection with the perspective of the ray will be the perspective of the shadow.

If *the shadow be on any vertical plane*, draw through the perspective of the horizontal projection of the point, a right line to the vanishing point of horizontal projections. It will be the perspective of the horizontal trace of a vertical plane of rays through the point. At the point where it intersects the perspective of the horizontal trace of the given plane erect a vertical line; it will be the perspective of the intersection of the plane of rays with the given plane. The point where this intersects the perspective of the ray through the given point, will be the perspective of the shadow.

281. The perspective of the indefinite shadow cast by a right line on a plane, may be constructed by finding the perspective of the point in which the line pierces the plane, and joining it with the perspective of the shadow cast by any other point of the right line; or by joining the perspectives of the shadows of any two points of the line by a right line.

If the line be of definite length join the perspectives of the shadows of its two extremities by a right line.

PRACTICAL PROBLEMS.

282. PROBLEM 70. *To construct the perspective of an upright rectangular pillar, with its shade and shadow on the horizontal plane.*

Let *lmno*, Fig. 118, be the lower base of the pillar in the horizontal plane, revolved as in Art. (269), and $p'q'$ the vertical projection of the upper base, s' *the principal point*, d_2 one of *the points of distance*, r_1 the vanishing point of rays, and r' of horizontal projections of rays. These important points will be thus represented in all the following problems.

Since *ml* and *no* are perpendicular to V, they vanish at s', Art. (265), and $m's'$ and $n's'$ are their indefinite perspectives; id_2 is the perspective of the diagonal *ni*, and *n* of the point *n*, Art. (268). n_1m_1, parallel to AB, is the perspective of *mn*, Art. (267); o_1 of the point *o*; n_1o_1 of the edge *no*, Art. (264); o_1l_1 of *ol*, and m_1l_1 of *ml*.

The edges of the upper base, which are horizontally projected in *on* and *lm*, pierce V at p' and q', and vanish at s'; hence $p's'$ and $q's'$ are their indefinite perspectives. The diagonal of the upper base, horizontally projected in *ni*, pierces V at i', $i'd_2$ is its perspective, and p_1 the perspective of the vertex of the upper base horizontally projected in *n*, and v_1 of that horizontally projected at *l*. The vertical edges which pierce H in *m*, *n*, *o*, and *l*, are parallel to V, and their perspectives parallel to themselves; hence m_1q_1, n_1p_1, o_1u_1, l_1v_1 are their perspectives terminating in the points q_1, p_1, u_1, v_1, and $q_1p_1u_1v_1$ is the perspective of the upper base.

The face of which $n_1o_1u_1p_1$ is the perspective is in the shade, and therefore is darkened in the drawing.

The shadow on H, of the edge represented by n_1p_1 is n_1p_2, Art (281). The shadow of the edge represented by p_1u_1 is parallel to the line itself, and therefore perpendicular to V, and vanishes at s'. It is limited at p_2 and u_2, Art. (281). u_2v_2, parallel to AB, is the perspective of the shadow of the edge represented by u_1v_1, and l_1v_2 of that represented by l_1v_1.

That part of the drawing within the line $n_1p_2.v_2x_1$ is darkened, being the perspective of that part of the shadow on H which is seen

283. Problem 71. *To construct the perspective of a square pyramid with its pedestal, and the perspective of its shadow.*

Let *mnol*, Fig. 119, be the base of the pedestal revolved, as in Art. (269), and *mnn'm'* its front face in the plane of the picture, *pqtu* the horizontal, and *p'u'* the vertical projection of the base of the pyramid, and C its vertex.

The face *mnn'm'* being in the plane of the picture is its own perspective. The four edges of the pedestal, which are perpendicular to V, pierce it in m, n, n' and m' and vanish at s'. The two diagonals *mo* and (mo, $m'n'$) pierce V at m and m' and vanish at d_1. The diagonals *nl* and (nl, $n'm'$) pierce V at n and n' and vanish at d_2. Hence o_1 is the perspective of o, Art. (268); k_1 of (on'); l_1 of l; and h_1 of (lm'); and the perspective of the pedestal is drawn as in Art. (282).

The two perpendiculars (ut, u') and (pq, p') pierce V at u' and p'; $u's'$ and $p's'$ are their perspectives intersecting $m'd_1$ and $n'd_2$ in u_1, q_1, p_1 and t_1 and $u_1p_1q_1t_1$ is the perspective of the base of the pyramid.

The perpendicular through C pierces V at c', and a diagonal through the same point at h' and $c's'$ and $h'd_1$ are their perspectives, and c_1 the perspective of C; and c_1u_1, c_1p_1, c_1q_1, and c_1t_1 are the perspectives of the edges of the pyramid.

nn_2 is the perspective of the shadow of nn' on H; n_2k_2 of the shadow of the edge represented by $n'k_1$ Art. (265). i_1 is the perspective of c, the horizontal projection of the vertex · i_1r' of the horizontal projection of a ray through C, and c_1r_1 the perspective of the ray; hence c_2 is the perspective of the shadow of C on H, Art. (280). e_1 is the perspective of the projection of C on the upper base of the pedestal, e_1r' of the projection of a ray on this plane, and c_3 of the shadow cast by C on this plane; and p_1c_3 the perspective of the shadow cast by the edge CP on this plane, Art. (281). This shadow passes from the upper base at a point of which x_1 is the perspective; x_1r_1 is the perspective of a ray through this point, intersecting n_2s' at x_2, the perspective of the shadow cast by one point of the edge CP on H, and x_2c_2 is the perspective of the shadow.

t_1c_3 is the perspective of the shadow cast by the edge CT on the upper base of the pedestal; y_1 of the point at which it passes from this base; y_2 of the shadow of this point on H, and y_2c_2 of the shadow of the edge CT.

The face represented by $c_1p\,q_1$ is in the shade, as also that represented by $nn'k_1o_1$, and both are darkened on the drawing.

$p_1x_1k\,y_1t_1$ bounds the darkened part on the perspective of the upper base, and $nn_2 \ldots c_2y_2 \ldots l_1$ that on H.

284. Problem 72. *To construct the perspective of a cylindrical column with its square pedestal and abacus, and also the shade of the column and shadow of the abacus on the column.*

Let $mngl$ (Fig. 120) be the horizontal projection of both pedestal and abacus, Art. (269); $mm'n'n$ the vertical projection of the pedestal, and $e'l'g'f'$ that of the abacus; $upqt$ being the horizontal projection of the column, $m'n'$ the vertical projection of its lower, and $e'f'$ of its upper base; the plane of the picture being coincident with that of the front faces of the pedestal and abacus.

Let the point of sight be taken as in Art. (273), in a plane through the centres of the upper and lower bases of the column and perpendicular to AB.

Construct the perspective of the pedestal as in Art. (282), $mm'n'n$ being its own perspective, and $m'l_1$ and $n'g_1$ the perspectives of those edges of the upper face of the pedestal which pierce V at m' and n'.

In the same way construct the perspective of the abacus: $e'l'g'f'$ being its own perspective, and $e's'$ and $f's'$ the indefinite perspectives of the edges of the lower face which pierce V at e' and f'.

$u_1t_1q_1p_1$ is the perspective of the lower base of the column determined as in Art. (273); u_1q_1 being its semi-conjugate and t_1p_1 its semi-tranverse axis. In the same way the perspective of the upper base is determined, its conjugate axis being the perspective of that diameter which pierces V at u'', horizontally projected in uq, and its tranverse axis z_1v_1 the perspective of the chord corresponding to tp.

t_1z_1 and p_1v_1 tangent to these ellipses, Art. (275), and perpendicular to AB, are the perspectives of the elements of contour of the column, and complete its perspective, Art. (274).

The elements of shade on the column are determined by two tangent planes of rays, Art. (251), and since these planes are vertical, their intersections with the plane of the upper face of the pedestal will be parallel to the horizontal projections of rays, and therefore vanish at r'. Since these intersections are also tangent to the lower circle of the column, their perspectives will be tangent to the ellipse $t_1u_1p_1$. Hence, if through r' two tangents be drawn to $t_1u_1p_1$, their points of contact will be points of the perspectives of the elements of shade. a_1k_2 is the perspective of the only one which is seen. The plane of rays by which this is determined intersects the lower edge $e'f'$ of the abacus in k_1, through which draw k_1r_1. It intersects a_1k_2 in k_2, the perspective of the point of shadow cast by k_1, and limiting the perspective of the element of shade.

Draw $r't_1$; it is the perspective of the intersection of a vertical plane of rays with the upper face of the pedestal. This plane intersects the column in an element represented by t_1z_1, and the edge represented by $e's'$ in a point of which y_1 is the perspective. Through y_1 draw y_1r_1 intersecting t_1z_1 in y_2, the perspective of the shadow cast on the column by the point Y, and in the same way the perspective of the shadow cast by any point of the same edge may be determined.

Through m' draw $m'r'$, and through e' draw $e'r_1$ intersecting the perspective of the element cut from the column in e_2, the perspective of the shadow cast by e' on the column, and in the same way the perspective of the shadow cast by any other point of the edge $e'f'$ may be found, as w_2 the perspective of the shadow cast by w''.

All of the lower face of the abacus is in the shade, and is darkened in the drawing. The line $y_2e_2..k_2$ is the perspective of the line of shadow on the column, and all above it is darkened, as all beyond a_1k_2. a_1r' is the perspective of the shadow cast by the element of shade on the upper surface of the pedestal, and the part beyond it which is seen is therefore darkened.

285. PROBLEM 73. *To construct the perspective of an upright cylindrical ring, with its shade and shadow on its interior surface.*

Let $mngl$, Fig. 121, be the horizontal projection of the outer cylinder of the ring, and ez_1h' its vertical projection; $upqt$ the horizontal, and $o_1i_1w_1$ the vertical projection of the inner cylinder, the plane of the picture being coincident with the plane of the front face of the ring.

The two circles ez_1h' and $o_1i_1w_1$ being in the plane of the picture are their own perspectives.

To construct the perspectives of the back circles horizontally projected in lg and tq, draw the vertical radius $(c, c'h')$ and the two vertical tangents at Q and G. h_1c, a_1q_1, and b_1g_1 are the perspectives of these lines, Art. (268), and c_1, q_1, and g_1 are the perspectives of the points C, Q, and G. With c_1 as a centre, and c_1q_1 and c_1g_1 as radii, describe the arcs t_1t'' and v_1g_1, they will be the perspectives of arcs of the back circles.

s_1y_1 and s_1x_1 are the perspectives of the elements of contour tangent to the circles ey_1x_1 and v_1g_1, Art. (275).

The element of shade on the outer cylinder is determined by a tangent plane of rays. This plane being perpendicular to V, its vertical trace will be parallel to r_1s', Art. (279), and tangent to y_1z_1h' at z_1, and z_1v_1 will be the perspective of this element.

Points of the shadow cast by the circle $o_1i_1w_1$ on the interior cylinder will be found by passing planes of rays perpendicular to the plane of the picture. Each plane will intersect the circle in a point which casts a shadow on the element which the plane cuts from the cylinder, Art. (252). o_1i_1 parallel to r_1s' is the vertical trace of such a plane, intersecting the circle in o_1, and the cylinder in an element of which i_1s' is the perspective. o_1r_1 is the perspective of a ray through o_1, and o_2 is the perspective of the shadow. The perspective of the shadow evidently begins at k_1.

286. **Problem** 74. *To construct the perspective of an inverted frustum of a cone with its shade and shadow.*

Let C, Fig. 122, be the vertex, *kolm* the horizontal, Art. (269), and $k'l'$ the vertical projection of the upper base, *ehgf* the horizontal, and $e'g'$ the vertical projection of the lower base, and $k'e'$ the vertical projection of one of the extreme elements.

The perspectives of the bases are determined as in Art. (273), $k_1o'y_1m_1$ of the upper, and $e_1h_1a_1f_1$ that of the lower base.

The perspective of the vertex is found as in Art. (268), CZ being the diagonal through it, piercing V at z'; and (co, c') the perpendicular piercing V at c', $z'd_2$ and $c's'$ are their perspectives intersecting at c_1. Through c_1 draw the two tangents c_1w_1 and c_1y_1; they are the perspectives of the elements of contour, and with the curves $k_1o'y_1m_1$ and $e_1h_1a_1f_1$ limit the perspective of the frustum.

The elements of shade on the cone are determined by two tangent planes of rays intersecting in a ray of light passing through the vertex, Art. (131), r_1c_1 is the perspective of this ray, and u_1r' the perspective of its horizontal projection, u_1 being the perspective of c; hence c_2 is the perspective of the point in which this ray pierces H, and the tangents c_2i_1 and c_2p_1 are the perspectives of the horizontal traces of the tangent planes, and i_1n_1 and

p_1q_1 of the elements of shade, Art. (253), the former only being seen.

That part of *the circumference of the upper base* towards the source of light, and between the points of which n_1 and q_1 are the perspectives, casts its shadow on the interior of the cone. Points of this shadow may be determined by intersecting the cone by planes of rays through the vertex. Each plane intersects the circumference in a point, and that part of the cone opposite the source of light in a rectilinear element. A ray of light through the point intersects the element in a point of the required shadow.

c_2a_1 is the perspective of the horizontal trace of such a plane. The plane intersects the cone in two elements represented by a_1y_1 and b_1c_1, and b_1 is the perspective of the point in which the plane intersects the circumference, b_1r_1 the perspective of a ray through this point, and b_2 the perspective of the required point of shadow.

The curve of shadow evidently begins at the points of which n_1 and q_1 are the perspectives.

The perspective of the lowest point of this shadow will be found by using the line c_2u_1, as this is the perspective of the trace of that plane which cuts out the element furthest from the point casting the shadow.

The perspective of the point of shadow on any element is found by using the line drawn through c_2 and the lower extremity of the perspective of the element. Thus the point of tangency b_2, Art. (275), is found by using c_2a_1 as the perspective of the trace of the auxiliary plane.

q_1k_1o' is the perspective of that part of the shadow on the interior which is seen, and is therefore darkened in the drawing as is the perspective of the shade $n_1i_1a_1y_1$.

i_1n_2 and p_1q_2 are the perspectives of the shadows cast by the elements of shade on H, the points n_2 and q_2 being determined by n_1r_1 and q_1r_1.

The plane determined by c_2a_1 intersects the circumference of the upper base in a point of which y_1 is the perspective, y_1r_1 is the perspective of a ray through this point piercing H at a point of which y_2 is the perspective. This is a point of the perspective of the curve of shadow of the upper circumference on H; and in the same way any number of points may be determined.

r_1v_2 drawn tangent to $n_1y_1m_1$ will also be tangent to $n_2y_2q_2$, since this tangent is the perspective of the element of contour of the cylinder of rays through the upper circumference by which its shadow is determined, Art. (275).

The curve $n_2y_2q_2$ is also tangent to i_1n_2 and p_1q_2 at n_2 and q_2.

287. PROBLEM 75. *To construct the perspective of a niche with its shadow on its interior surface.*

Let the niche be formed by a semi-cylinder, the lower base of which is mlo, Fig. 123, and the upper base vertically projected in $m'o'$, and the quarter sphere vertically projected in the semicircle $m'k'o'$, its lower semicircle being coincident with the upper base of the cylinder, and the plane of the picture so taken as to contain the elements mm' and oo' and the front circle $m'k'o'$.

These elements and front circle being in the plane of the picture will be their own perspectives.

An arc of an ellipse, ml_1o, is the perspective of the lower base of the cylindrical part, and $m'n_1o'$ that of the upper base, these being constructed as in Art. (273), l_1 being the perspective of l and n_1 of the corresponding point of the upper base. These lines form the perspective of the niche.

The lines which cast shadows on the interior of the niche are the element mm' and the arc $m'k'e'$.

The shadow cast by mm' is determined by passing through it vertical plane of rays; mr' is the perspective of its trace, Art. (277). This plane cuts from the cylinder an element represented by f_2m_2, the intersection of which by $m'r_1$ in m_2 is the perspective of the shadow cast by m' on the element; and f_2m_2 is the perspective of the shadow cast by $m'f'$, a part of mm', on the interior of the cylinder; mf_2 is the perspective of the shadow cast by mf' on H.

Points of the shadow cast by the front circumference, $m'k'e'$, *on the cylindrical part of the niche* may be determined by intersecting the cylinder by vertical planes of rays. Each plane intersects the cylinder in a rectilinear element and the arc in a point. A ray of light through this point intersects the element in a point of the curve of shadow. $r'g$ is the perspective of the horizontal trace of such a plane. It intersects $m'k'e'$ in g', and the cylinder in an element represented by i_1g_2 and g_2 is the perspective of the shadow cast by g' on the element.

The shadow cast by a part of the front circumference on *the spherical part of the niche* is an equal arc of a great circle. For this shadow is determined by a cylinder of rays through the front circumference, and the part of each element of the cylinder between the point casting the shadow and its shadow is a chord of the sphere, and all these chords will be bisected by a plane through c' perpendicular to them. These half chords may be regarded, one set as the ordinates of the arc casting the shadow, and the other as the ordinates of the shadow, and since the corresponding ordinates of the two arcs are equal, and in all respects like situated, the two arcs will be equal. Their planes evidently intersect in the radius $c'e'$ perpendicular to the axis of the cylinder of rays, or to the ray of light, and also perpendicular to the projection $s'r_1$ of the ray of light on the plane of the picture, Art. (276).

Points of this shadow may be constructed by intersecting the quarter sphere by planes parallel to the plane of the picture. Each plane will cut from the quarter sphere a semi-circumference. The front semi-circumference will cast upon this plane a shadow parallel and equal to itself, Art. (246), and the intersection of this with the semi-circumference cut from the sphere will be a point of the required shadow, Art. (244).

Draw y_1z_1 parallel to $m'o'$, it may be taken as the perspective of the intersection of an auxiliary plane with the upper base of the cylinder, and the semi-circumference $y_1x_2z_1$ is the perspective of the semi-circumference cut from the sphere.

The centre of the front circle casts a shadow upon this plane whose perspective is c_2, Art. (280), c'' being the perspective of the projection of c' on this plane, and $c''c_2$ parallel to $s'r_1$ the perspective of the projection of the ray through c' on this plane. c_2m_3 parallel to $c'm'$ is the perspective of the shadow cast by the radius $c'm'$ on this plane, Art. (267), and m_3x_2 the perspective of the arc of shadow intersecting $y_1x_2z_1$ in x_2, a point of the perspective of the shadow.

Otherwise thus: Draw $k'k''$ parallel to $s'r_1$, it may be taken as the vertical trace of a plane of rays perpendicular to V. This plane cuts from the quarter sphere a semi-circumference. A ray of light through k' will intersect this semi-circumference in a point of the curve of shadow. Revolve the plane about $k'k''$ until it coincides with V. $k'x''k''$ will be the revolved position of the semi-circumference. Revolve the ray of light through the point of sight about $s'r_1$ until it coincides with V; $s''r_1$ will be its revolved position, $s's''$ being equal to $s'd_1$, Art. (266). Through k' draw $k'x''$ parallel to $s''r_1$, it will be the revolved position of the ray through k', and x'' will be the revolved position of a point of the shadow. Through x' draw $x's'$; it is the perspective of the perpendicular $x'x''$ intersecting $k'r_1$ in x_2, the perspective of the shadow cast by k'.

The perspective of this curve of shadow evidently begins at e', the point of tangency of a line parallel to $s'r_1$.

The point at which the curve of shadow passes from the spherical to the cylindrical part is evidently the point in which the intersection of the plane of the curve of shadow with the plane of the upper base meets the circumference of the upper base. To determine the perspective of this point, through x_2 draw x_2u_1 parallel to $c'e'$; it is the perspective of a line of the plane of the curve of shadow as also of the plane of the circle of which m_3x_2 is the perspective. u_1 is the perspective of the point where this line pierces the plane of the upper base, and $c'u_1$ the perspective of the intersection of the plane of the circle of shadow with the plane of the upper base, and $p_{\prime}$ the perspective of the required point.

288. PROBLEM 76. *To construct the perspective of a sphere with its shade, and shadow on the horizontal plane.*

Let the plane of the picture be taken perpendicular to the right line, joining the point of sight with the centre of the sphere, and tangent to the sphere; then, Fig. 124, c will be the horizontal and s' the vertical projection of the centre, ca being equal to the radius and $b'c''c'''$ the vertical projection of the sphere.

The perspective of the sphere will be determined by a visual cone tangent to it, and the circle in which this cone is intersected by the plane of the picture will be the outer line of the perspective, Art. (274).

To construct this, pass a plane through S and C perpendicular to AB. It intersects the sphere in a great circle, and the cone in rectilinear elements tangent to this circle and piercing V in points of the required circumference. $am_{\prime}$ is the vertical trace of this auxiliary plane. Revolve the plane about $am_{\prime}$ as an axis until it coincides with V; S, in revolved position, will be at $d_{\prime}$ and C at c'', and $m's'$ described with the centre c'' and radius $c''s'$ will be the revolved position of an arc of the circle cut from the sphere, and $m'd_{\prime}$ that of one of the tangents. In true position this tangent pierces V at $m_{\prime}$, and the circle $m_{\prime}u_{\prime}v_{\prime}$ described with s' as a centre and $s'm_{\prime}$ as a radius will be the perspective of the sphere.

The curve of shade on the sphere, the line of contact of a tangent cylinder of rays, is the circumference of a great circle whose plane is perpendicular to the axis, Art. (121), or to the ray of light.

The plane of rays through the point of sight whose vertical trace is $s'r_{\prime}$, is perpendicular to the plane of the circle of shade, and is evidently a principal plane, Art. (206), of the visual cone by which the perspective of this curve is determined, and since the plane of the picture is perpendicular to this plane, its inter section with the cone will be an ellipse, all the chords of which perpendicular to $s'r_1$ are bisected by it. This line will then be an indefinite axis of the ellipse. *Ana. Geo.*, Art. (85). This princi pal plane intersects the circle of shade in a diameter whose perspective will be this axis.

To determine it, revolve the plane about $s'r_1$ until it coincides with V. S is found at s'', $s's''$ being equal to $s'd_1$. The centre of the circle cut from the sphere will be at c''', $n'l'p'$ will be the revolved position of the circle, and $n'p'$ perpendicular to $s''r_1$ will be the revolved position of the diameter. Two visual rays from its extremities pierce V at n_1 and p_1, and n_1p_1 is its perspective and the required axis.

o_1, the middle point of this axis, is the centre of the ellipse, and l_1k_1 perpendicular to n_1p_1, the position of the other axis, and o' the revolved position of the point of which o_1 is the perspective. If through l_1k_1 and S a plane be passed, it will intersect the circle of shade in a chord passing through the point of which o' is the revolved position, perpendicular to the plane of rays of which $s'r_1$ is the trace and equal to $l'k'$. The perspective of this chord will be the other axis.

To determine it, revolve this plane about k_1l_1 until it coincides with V. S will be found at s''', o_1s''' being equal to o_1s''; o' will be at o'', o_1o'' being equal to o_1o', and $k''l''$ perpendicular to $s'r_1$ will be the revolved position of the chord, the perspective of which is k_1l_1, the other axis.

The ellipse described upon these two axes is the perspective of the curve of shade. It is tangent to $m_1u_1v_1$ at u_1 and v_1, since these are the perspectives of two points which are determined by two visual planes of rays tangent to the sphere, whose traces are r_1u_1 and r_1v_1. These are the perspectives of the points in which the curve of shade crosses the apparent contour of the sphere, Art (275).

The curve of shadow on the horizontal plane is cast by the curve of shade. c_2 is the perspective of the shadow cast by the centre,

c_1 being the perspective of its horizontal projection, and $c\,r'$ the perspective of the horizontal projection of a ray through the centre, Art. (280).

The perspective of the shadow of each diameter of the curve of shade must pass through c_2. If, then, we find the perspective of the point in which a diameter pierces the horizontal plane and join it with c_2, we shall have the indefinite perspective of the shadow of this diameter, and the perspectives of two rays through its extremities will intersect the perspective of this shadow in points of the perspective of the curve of shadow.

All diameters of the circle of shade pierce the horizontal plane in the horizontal trace of the plane of this circle. This trace being a horizontal line must vanish in the horizon of the picture $s'd$, Art. (267), and being in the plane of the circle of shade must also vanish in the vanishing line of this plane; and hence the intersection of these two lines will be a point of the perspective of the horizontal trace.

To construct the vanishing line of the plane of shade, through s'' draw $s''x'$ perpendicular to $s''r_1$; it is the revolved position of a line passing through S in the plane of rays whose vertical trace is r_1s' and parallel to the plane of shade. It pierces V at x', one point of the vertical trace of a plane through S parallel to the circle of shade. Through x' draw $x'q_1$ perpendicular to $x'r_1$, Art. (43); it is the vertical trace, or required vanishing line. This intersects $s'q_1$ at q_1.

To determine another point of the perspective of the horizontal trace, find the perspective of the point in which that diameter of the circle of shade, which is parallel to the plane of the picture, pierces H. The horizontal projection of this diameter is parallel to AB; hence, c_1i_1 is its perspective. $s's''$ perpendicular to r_1s' is the perspective of the diameter itself, for these two lines are the orthographic projections of the diameter and ray of light through the centre, since the visual plane through each of them contains the projecting perpendicular Ss', Art. (36). i_1 is then the perspec tive of the point in which this diameter pierces H, and q_1i_1 the perspective of the horizontal trace.

Join i_1 with c_2; i_1c_2 is the perspective of the shadow cast by the diameter represented by g_1f_1. Join g_1 and f_1 with r_1 by right lines. These are the perspectives of rays through the extremities of the diameter, and g_2 and f_2 are the perspectives of points of the curve of shadow.

Produce u_1e_1 to y_1, y_1 is the perspective of the point in which the corresponding diameter pierces H, and y_1c_2 the perspective of its shadow, and e_2 and u_2 the perspectives of the shadows cast by its extremities. In the same way other points of the curve are constructed.

The curve must be drawn tangent to r_1u_1 and r_1v_1 at u_2 and v_2, since these are the perspectives of the points in which the curve of shadow crosses the elements of apparent contour of the cylinder of rays, by which the curve of shadow is determined.

289. Problem 77. *To construct the perspective of a groined arch with its shadows.*

To form the groined arch here considered, two equal right semi-cylinders, with semi-circular bases, are so placed that their axes shall be at right angles and bisect each other and the diameters of the semi-circular bases in the same horizontal plane. Thus, Fig. 125, let *nolu* be the horizontal, and *ok'l* the vertical projection of one semi-cylinder, and *fghi* the horizontal, and *pp'q'q* the vertical projection of the second cylinder, *pp'* and *qq'* being the vertical projections of the two semi-circular bases.

These two cylinders intersect in two equal ellipses horizontally projected in the diagonals *ab* and *ce*. These ellipses are *the groins*. The arch is now formed by taking out from each cylinder that part of the other which lies within it. Thus, all that part of the cylinder whose elements are parallel to V, horizontally projected in *akc* and *bke*, is removed, as also that part of the other cylinder projected in *ake* and *bkc*.

The arch thus formed is placed upon four pillars standing in the four corners of a square, and whose horizontal projections are the four squares *mnai*, *uchv*, etc. The elements *ih* and *fg* being coincident with the inner upper edges of the pillars which are horizontally projected in *ia-ch* and *fe-bg*, and the elements *no* and *ul* coincident with the inner upper edges which are projected in *na-eo* and *uc-bl*.

In this position of the arch the front circumference, or front base of the cylinder whose elements are perpendicular to V springs from the upper extremities of the two edges of the front pillars which are horizontally projected at *n* and *u* and is described on a diameter equal and parallel to *nu*. The back circumference, or other base of the same cylinder, springs from the upper extremities of the edges horizontally projected at *o* and *l*.

The side circumferences, or bases of the other cylinder, spring, the one from the upper ends of the two edges horizontally projected in *i* and *f*, and the other from those projected in *h* and *g*, and their diameters are parallel and equal to *if* and *hg*.

The groins spring from the upper ends of the four inner edges of the pillars, the one from those projected in *a* and *b*, and the other from those projected in *c* and *e*, and *ab* and *ce* are respectively equal and parallel to the transverse axes of the groins, the common conjugate axis being equal and parallel to the radius $k'k''$.

The planes of the outer faces of the pillars are produced upwards, inclosing the mass of masonry supported by the arch.

To construct the perspective of the arch thus placed let us take, Fig. 126, the plane of the picture coincident with the plane of the front faces of the front pillars, *mm'n'n* and *uu'v'v* being these front faces and their own perspectives, and let the principal point be at s' in a plane perpendicular to V and midway between the pillars.

The perspectives of the front pillars are constructed precisely as in Art. (282); a_1 and i_1 being the perspectives of the upper ends of the two back edges of the first, and c_1 and h_1 those of the second pillar; md_1, the perspective of the diagonal corresponding to mq in Fig. 125, intersects us', the perspective of the perpendicular corresponding to ul, in b_2, which is the perspective of the lower end of the edge of the back pillar corresponding to b; b_2g_2, parallel to AB, is the perspective of the edge corresponding to bg, and b_2l_2 that corrresponding to bl. Through b_2, g_2, and l_2 draw b_2b_1, g_2g_1, and l_2l_1 until they meet $u's'$ and $v's'$ in b_1, g_1, and l_1, and draw b_1g_1 and b_1l_1, thus completing the perspectives of the two faces of the back pillars which can be seen.

In the same way the perspective of the other back pillar may be constructed, e_2 being the perspective of the point corresponding to e.

On $n'u'$ as a diameter describe the semicircle $n'k'u'$; it is the front semicircle of the arch in the plane of the picture and its own perspective. The other base of the same cylinder being parallel to the plane of the picture will be in perspective a semicircle, and since it springs from the points of which o_1 and l_1 are the perspectives, o_1l_1 will be the diameter and the semicircle on this its perspective.

To obtain points of *the perspectives of the groins and side circles*, intersect the arch by horizontal planes. Each plane will cut from the cylinders rectilinear elements, the intersection of which will be points of the groins; and from the side faces of the arch, right lines which will intersect the elements parallel to the plane of the picture in points of the side circles.

Let $m''v''$ be the trace of such a plane. It intersects the cylinder whose axis is perpendicular to V in two elements which pierce V at w' and z', vanish at s', and are represented by $w's'$ and $z's'$. The same plane intersects the side faces of the arch in lines represented by $m''s'$ and $v''s'$. The two diagonals through m'' and v'', intersect the elements in points of the groins, see Fig. 125. $m''d_1$ and $v''d_2$ are the perspectives of these diagonals intersecting $w's'$ in w'' and w''' and $z's'$ in z'' and z''', points of the perspectives of the groins. $w''z''$ and $w'''z'''$ are the perspectives of the elements cut from the cylinder whose axis is parallel to V. These are intersected by $m''s'$ and $v''s'$ in points p_1 and q_1 of the perspectives of the side circles.

The perspectives of the groins thus determined are drawn, one from a_1 to b_1, the other from c_1 to e_1, and the perspectives of the side circles, one from i_1 to f_1, and the other from h_1 to g_1.

Since all four of these curves cross the element of contour of the cylinder whose axis is parallel to the plane of the picture, their perspectives will be tangent to the perspective of this element. To determine it, through S pass a plane perpendicular to the axis of the cylinder, r_2s' is its trace. It cuts from the cylinder a circle and from the visual plane tangent to the cylinder a right line tangent to the circle. Revolve this plane about r_2s' until it coincides with V. The centre of the circle will be at m', and S at d_1. The semicircle described with m' as a centre and $k''n'$ as a radius is the revolved position of the circle cut from the cylinder, and the tangent to it from d_1 the revolved position of the line cut from the tangent plane. This pierces V at x', a point of the perspective of the element of contour. The line through x' parallel to AB, is the tangent to all the curves.

nr' and i_2r' are the perspectives of the indefinite shadows cast on H by the vertical edges of the left hand pillar, Art. (282). The point in which nr' intersects b_2g_2 is the perspective of a point of the shadow of nn' on the front face of the back pillar. This line being parallel to this face, its shadow will be parallel to itself as also the perspective of the shadow, which is terminated at n_2, the perspective of the shadow cast by n', Art. (280).

From this point the shadow is cast on this face by the front circle, 'and is the arc of a circle as is its perspective, Art. (272). The shadow of the radius $n'k''$ is parallel to itself, as also the perspective of this shadow passing through n_2 and terminated at k_2. The arc described with x_2 as a centre and k_2n_2 as a radius is then *the perspective of the shadow of the front circle on the face.*

e_2r' is the perspective of the shadow cast on H by the edge represented by e_2e_1, a small part of which only lies in the limits of the drawing.

Points of *the shadow cast by the front circle on the cylinder*, of which it is the base, are determined by intersecting the cylinder by planes of rays perpendicular to the plane of the picture. The traces of these planes are parallel to $s'r_1$, Art. (279), and each plane cuts the front circumference in a point casting the shadow and the cylinder in a rectilinear element, which is intersected by the ray through the point in its shadow.

Let $y\,y''$ be the trace of such a plane. It intersects the cylinder in an element represented by $y''s'$, and the circumference in the point y', and y_2 is the perspective of the shadow. The perspective of the shadow begins at the point in which a trace parallel to $y''y'$ is tangent to the circle.

Points of *the shadow cast by the side circle on the left*, on the cylinder of which it is a base, may be found by intersecting the cylinder by planes of rays perpendicular to the side faces of the arch. Each plane intersects the circumference in a point casting the shadow and the cylinder in a rectilinear element which is intersected by the ray through the point in a point of the shadow.

The intersections of these planes with the outer side face of the arch vanish at r_2, Art. (278). Let r_2t_1, be the perspective of one of these intersections. The plane intersects the side circumference in a point of which t_1 is the perspective and the cylinder in an element represented by p_1q, and t_2 is the perspective of the point of shadow. The perspective of the curve evidently begins at the point of tangency of a line through r_2 tangent to $i_1p_1f_1$.

290. Although in general the perspectives of objects are made as in the preceding articles on a plane, it is not unusual to make them upon curved surfaces.

Extended *panoramic views* are thus made upon a right cylinder with a circular base, the point of sight being in the axis, and the observer thus entirely surrounded by pictures of objects.

Objects are also sometimes represented on the interior of a spherical dome.

In all cases the perspectives of points are constructed by drawing through the points visual rays, and finding the points in which these pierce the surface on which the representation is made.

The constructions involve only the principles contained in Arts (263), (264), and (268).

PART V.

ISOMETRIC PROJECTIONS.

PRELIMINARY DEFINITIONS AND PRINCIPLES.

291. Let three right lines be drawn intersecting in a common point and perpendicular to each other, two of them being horizontal and the third vertical; as the three rectangular co-ordinate axes in space, *Ana. Geo.*, Art. (42), or the three adjacent edges of a cube; then let a fourth right line be drawn through the same point, making equal angles with the first three, as the diagonal of a cube. If a plane be now passed perpendicular to this fourth line, and the right lines and other objects be orthographically projected upon it, the projections are called *Isometric.*

The three right lines first drawn are *the co-ordinate axes;* and the planes of these axes, taken two and two, are *the co-ordinate planes.* The common point is *the origin.*

The fourth right line is *the Isometric axis.*

If A designate the origin, the co-ordinate axes are designated, as in *Ana. Geo.*, as the axes AX, AY, and AZ, the latter being vertical; and the co-ordinates planes as the plane XY, XZ and YZ, the first being horizontal, and the other two vertical.

292. Since the co-ordinate axes make equal angles with each other and with the plane of projection, it is evident that their pro-

jections will make equal angles with each other, two and two, that is, angles of 120°. Hence, Fig. 128, if any three right lines, as Ax, Ay, and Az, be drawn through a point as A, making with each other angles of 120°, these may be taken as *the projections of the co-ordinate axes*, and are *the directrices* of the drawing.

It is further evident, that if any equal distances be taken on the co-ordinate axes, or on lines parallel to either of them, their projections will be equal to each other, since each projection will be equal to the distance itself into the cosine of the angle of inclination of the axes with the plane of projection, Art. (197).

The angle which the diagonal of a cube makes with either adjacent edge is known to be 54° 44′; therefore the angle which either edge, or either of the co-ordinate axes, makes with the plane of projection will be the complement of this angle, viz., 35° 16′.

293. If a scale of equal parts, Fig. 127, be constructed, the unit of the scale being *the projection* of any definite part of either co-ordinate axis, as *one inch*, *one foot*, etc., that is, *one inch multiplied by the natural cosine of* 35° 16′=.81647 of an inch; we may, from this scale, determine the true length of the isometric projection of any given portion of either of the co-ordinate axes, or of lines parallel to them, by taking from the scale the same number of units as the number of inches, or feet, etc., in the given distance. *Conversely*, the true length of any corresponding line in space may be found by applying its projection to the isometric scale, and taking the same number of inches, or feet, etc., as the number of parts covered on the scale.

294. Since in most of the frame work connected with machinery, and in the various kinds of buildings, the principal lines to be represented occupy a position similar to that of the co-ordinate axes, viz., perpendicular to each other, one system being ver-

tical, and two others horizontal, the isometric projection is used to great advantage in their representation.

A still greater advantage arises from the fact that in a drawing thus made, all lines parallel to the directrices are constructed on the same scale, Art. (292).

ISOMETRIC PROJECTIONS OF POINTS AND LINES.

295. If a point be given, as in Analytical Geometry, by its co-ordinates, or its three distances from the co-ordinate planes, Art. (40), its isometric projection may be easily constructed. Thus, Fig. 128, let Ax, Ay, and Az be the directrices, A being the projection of the origin. On Ax lay off Ap, equal to the same number of units, *taken from the isometric scale*, as there are units in the distance of the point from the co-ordinate plane YZ. Through p draw pm' parallel to Ay, and make it equal to the number of units in the distance of the point from the plane XZ. Through m' draw $m'm$ parallel to Az, and make it equal to the third given distance, and m will be the required projection.

296. The projection of any right line parallel to either of the co-ordinate axes may be constructed by finding, as above, the projection of one of its points, and drawing through this a line parallel to the proper directrix.

If the line is parallel to neither of the axes, the projections of two of its points may be found as above and joined by a right line.

The projections of curves may be constructed by finding a sufficient number of the projections of their points.

297. If the circumference of a circle be in, or parallel to, either co-ordinate plane, its projection may be constructed thus: At the

extremities of the two diameters, which are parallel to the co-ordinate axes, draw tangents, thus circumscribing the circle by a square. The projections of each set of tangents will be parallel to the projection of the parallel diameter and tangent to the projection of the circle, Art. (65); hence the projections of these two equal diameters will be equal conjugate diameters of the ellipse which is the projection of the circle, and since these projections are parallel to the directrices they will make with each other an angle of 120°. Upon these the ellipse may be described, taking care to make it tangent to the projections of the four tangents.

PRACTICAL PROBLEMS.

298. PROBLEM 78. *To construct the isometric projection of a cube.*

Let the origin be taken at one of the upper corners of the cube, the base being horizontal, and let Ax, Ay, and Az, Fig. 129, be the directrices.

From A, on the directrices, lay off the distances Ax, Ay, and Az, each equal to the same number of units, taken from the isometric scale, as there are units of length in the edge of the cube. These will be the projections of the three edges of the cube which intersect at A. Through x, y, and z draw xe, xg, ye, yc, zc, and zg parallel to the directrices, completing the three equal rhombuses, Axey, etc. These will be the projections of the three faces of the cube which are seen, and the representation will be complete.

299. The ellipses constructed upon the equal lines kl, qs; st uv, &c., Fig. 129, are evidently the projections of three circles inscribed in the squares which form the faces of the cube, and these lines are the equal conjugate diameters of the ellipses, Art. (297) lk being drawn through the middle point of Ax, and sq through

the middle point of Az, and parallel respectively to Az and Ax, m being the projection of the centre of the circle.

300. PROBLEM 79. *To construct the isometric projection of an upright rectangular beam with its shade and shadow on the horizontal plane.*

Let the origin A and the directrices, Fig. 130, be taken as in the preceding problem.

On Ax lay off the distance Ax equal to the breadth of the beam in units taken from the isometric scale; on Ay, its thickness, and on Az, its length, and complete the three parallelograms Axcz, Aybz, and Axey, These will be the projections of the three faces of the beam which are seen.

Let a' be assumed as the isometric projection of the point in which a ray of light through A pierces H; then Aa' will be the isometric projection of the ray, and za' the projection of the shadow of the edge AZ, Art. (248). Through a' draw $a'y'$ parallel and equal to Ay. It is the projection of the shadow of the edge AY. $y'e'$ equal and parallel to ye is the projection of the shadow cast by YE, and de' that of the edge DE.

The face represented by Azby is in the shade.

301. PROBLEM 80. *To construct the isometric projection of the framework of a simple horizontal platform resting on four rectangular supports.*

Let A, Fig. 131, be the upper corner of one of the horizontal beams, the directrices being as in the preceding problems.

Lay off Ax, Ab, and Az equal respectively to the number of units in the length, breadth, and thickness of the horizontal beam, and complete the three parallelograms Aq, Ax', Ab', thus forming the representation of the beam. Lay off zc equal to the distance of the first support from the end of the beam, cd equal to its thickness, cc' parallel to Az equal to its height, and $c'e'$ parallel to

Ay its breadth, and complete the parallelograms dc' and ce'. Lay off zf equal to the distance of the second support from the end of the beam, and complete its projection in the same way.

Lay off ba equal to the horizontal distance between the two beams, and construct the projection of the second beam and its supports precisely as the first was constructed.

From a to n lay off the distance from the end of the beam to the first cross-piece. Make nn' equal to its breadth, and np parallel to Az equal to its thickness, and draw lines through n, n', and p parallel to Ay. Lay off ar equal to the distance of the second cross-piece from the end of the beam, and construct its projection in the same way as the first.

The faces of the frame-work in the shade and seen are darkened in the drawing.

302. PROBLEM 81. *To construct the isometric projection of a house with a projecting roof.*

Let A, Fig. 132, be the upper end of the intersection of the front and side faces of the house, Ax, Ay, and Az the directrices.

On Ax lay off the length, on Ay the breadth of the house, and on Az the height of the side walls, and complete the two parallelograms Ap and Ap', they are the projections of the side and front faces of the house.

At b, the middle point of Ay, draw br parallel to Az, and make it equal to the height of the ridge of the roof above AY, and join rA and ry; these lines will be the projections of the intersections of the face ZY with the roof. From r lay off rr' equal to the distance that the roof projects beyond the walls of the house, and draw $r's$ and $r's'$ parallel respectively to rA and ry. On Ay lay off Ad and ye each equal to rr', and draw dd' and ee' parallel to Az, intersecting Ar and ry produced in d' and e'. These will be points of the projections of the eaves. Draw $r'r''$ parallel to Ax, and make it equal to the length of the ridge, and complete the parallelograms $r'i$ and $r't'$. These will be the projections of the two inclined faces of the roof.

From b lay off bc and bf, each equal to the distance of the side faces of the chimney from the ridge, and draw cg and fg' parallel to Az, intersecting Ar and ry in g and g', and draw gh and $g'h'$ parallel to $r'r''$. They will be the projections of the intersections of the planes of the side faces of the chimney with the roof. Make gh equal to the distance of the front face of the chimney from the front face of the house, and draw hi and ih' parallel to Ar and ry; they will be the projections of the intersection of the front face of the chimney with the roof. Make hk equal to the thickness of the chimney, and hl equal to its height, and complete its projection as in the Figure.

From z lay off zu equal to the distance of the door from the edge Az, uu' equal to its width, and uw equal to its height, and complete the parallelogram $u'w$. Make zq equal to the distance of the windows above the base, qn and qo their distances from the edge, Az, nn', and oo' their width, and nm equal to their height, and complete the parallelograms.

303. A knowledge of the preceding simple principles and constructions will enable the draughtsman to make isometric drawings of the most complicated pieces of machinery, and most extended collections of buildings, walls, etc.; drawings which not only present to the eye of the observer a very good representation of the objects projected, but are of great use to the machinist and builder.

Assume any 3 right lines not in the same plane. " point in one of them. Construct a plane through this point and one of the other lines. Find where the third line pierces this plane. Join this point with the assumed first point by a right line. Construct planes perpendicular to the planes of this 4th line and either two of the original first assumed lines at the point of their intersection & bisecting the angle between them. Find the intersection of the perpendicular planes. It will be the axis of a Hyperboloid of one Revolution of one nappe in which the three first assumed lines lie.

Check. Construct a third plane perpendicular to the 4th line & the 3d assumed line at their intersection & bisecting the angle between them this should pass through the intersection of the other two perpendicular planes.

www.ingramcontent.com/pod-product-compliance
Lightning Source LLC
LaVergne TN
LVHW011217110826
845150LV00006B/1449

9781425516772